Phytochemistry of Plants of Genus *Rauvolfia*

Phytochemical Investigations of Medicinal Plants

Series Editor:
Brijesh Kumar

Phytochemistry of Plants of Genus *Phyllanthus*
Brijesh Kumar, Sunil Kumar and K. P. Madhusudanan

Phytochemistry of Plants of Genus *Ocimum*
Brijesh Kumar, Vikas Bajpai, Surabhi Tiwari and Renu Pandey

Phytochemistry of Plants of Genus *Piper*
Brijesh Kumar, Surabhi Tiwari, Vikas Bajpai and Bikarma Singh

Phytochemistry of *Tinospora cordifolia*
Brijesh Kumar, Vikas Bajpai and Nikhil Kumar

Phytochemistry of Plants of Genus *Rauvolfia*
Brijesh Kumar, Sunil Kumar, Vikas Bajpai and K. P. Madhusudanan

Phytochemistry of *Piper betle* Landraces
Vikas Bajpai, Nikhil Kumar and Brijesh Kumar

For more information about this series, please visit: https://www.crcpress.com/ Phytochemical-Investigations-of-Medicinal-Plants/book-series/PHYTO

Phytochemistry of Plants of Genus *Rauvolfia*

Brijesh Kumar, Sunil Kumar,
Vikas Bajpai and K. P. Madhusudanan

CRC Press
Taylor & Francis Group
Boca Raton London New York

CRC Press is an imprint of the
Taylor & Francis Group, an **informa** business

First edition published 2020
by CRC Press
6000 Broken Sound Parkway NW, Suite 300, Boca Raton, FL 33487-2742

and by CRC Press
2 Park Square, Milton Park, Abingdon, Oxon, OX14 4RN

ISBN: 978-0-367-85752-3 (hbk)
ISBN: 978-1-003-01484-3 (ebk)

Typeset in Times
by codeMantra

Contents

List of Figures

List of Tables

List of Scheme

Preface

Herbal medicines have been widely used for hundreds of years all over the world. Plant constituents are used directly either as therapeutic agents or as starting materials for searching lead molecules for the synthesis of drugs or as models for pharmacologically active compounds. Natural products have played a significant role in modern drug discovery, and many drugs in market were discovered from natural sources. The popularity of herbal medicine can be attributed to a number of factors. Herbal medicines are cheap, of natural origin and have no side effects. The most important challenges faced by herbal medicines and formulations arise because of their lack of complete standardization. The efficiencies of herbal medicines depend on the amount of active components present in them, which could vary significantly in contents. Therefore, quality control (QC) of herbal medicines becomes a very important issue. The discovery of relevant metabolites and fingerprints allow the introduction of an appropriate QA/QC in traditional medicine formulations. Emergence of very robust technologies in modern analytical methods has led to a reliable and fast analysis of medicinal plants.

Rauvolfia species, commonly known as Sarpagandha and snakeroot, has been traditionally used in Ayurveda for treating high blood pressure, hypertension, snake bites, fever and insanity. The scientific studies and research data available on *Rauvolfia* species demonstrate their enormous potential. Due to its wide variety and differences in chemical composition, it is necessary to develop an efficient and reliable method for rapid screening and determination of phytochemical constituents in root extracts of plants of *Rauvolfia* genus. In this book, we have included an ethno- and phytopharmacological review and described a new approach for the development and validation of rapid, sensitive and reliable analytical methods, namely, using HPLC-QTOF-MS/MS and UPLC-QqQLIT-MS/MS, for comparative screening and determination of monoterpene indole alkaloids in ethanolic extracts of six *Rauvolfia* species found in India.

31/12/2019   **The Author**

Acknowledgments

The completion of this book is due to the Almighty who blessed us with all the resources required to accomplish this journey and the warmest support and helpful advice of many colleagues whom we owe so much. We are glad to express our gratitude to people who have been supportive to us at every step. We express our deep sense of gratitude to Sophisticated Analytical Instrument Facility (SAIF) Division, CSIR-Central Drug Research Institute (CDRI), Lucknow, India, for their support. All the team members thank the Director of CSIR-CDRI for his support during this period.

Authors

Dr. Brijesh Kumar is a Professor (AcSIR) and Chief Scientist of Sophisticated Analytical Instrument Facility (SAIF) Division, CSIR-Central Drug Research Institute (CDRI), Lucknow, India. Currently, he is facility in charge at SAIF Division of CSIR-CDRI. He has completed his PhD from CSIR-CDRI, Lucknow (Dr. R.M.L Avadh University, Faizabad, UP, India). He has to his credit 7 book chapters, 1 book and 145 papers in reputed international journals. His current area of research includes applications of mass spectrometry (DART MS/Q-TOF LC-MS/4000 QTrap LC-MS/ Orbitrap MS^n) for qualitative and quantitative analyses of molecules for quality control and authentication/standardization of Indian medicinal plants/parts and their herbal formulations. He is also involved in the identification of marker compounds using statistical software to check adulteration/substitution.

Dr. Sunil Kumar is currently working as an Assistant Professor at Ma. Kanshiram Government Degree College, Farrukhabad, Uttar Pradesh, India. He has completed his PhD under the supervision of Dr. Brijesh Kumar on application of mass spectrometric techniques in qualitative and quantitative analyses of phytoconstituents and identification of chemical markers by chemometric technique in *Phyllanthus* and *Rauvolfia* Spp. in SAIF at CSIR-CDRI, Lucknow, India. His research interest includes qualitative and quantitative analyses of phytochemicals using LC-MS/MS analysis.

Dr. Vikas Bajpai completed his PhD from the Academy of Scientific and Innovative Research (AcSIR), New Delhi, India, and carried out his research work under the supervision of Dr. Brijesh Kumar at CSIR-CDRI, Lucknow, India. His research interest includes the development and validation of LC-MS/MS methods for qualitative and quantitative analyses of phytochemicals in Indian medicinal plants.

Dr. K. P. Madhusudanan is a mass spectrometry scientist born in 1947 in Kerala, India. He obtained his doctoral degree in 1975 specializing in organic mass spectrometry in National Chemical Laboratory, Pune, India. He worked as a Scientist and Head of SAIF in CSIR-CDRI, Lucknow, India, until 2007. His research experience since 1970 includes various aspects of organic mass spectrometry such as fragmentation mechanism, gas phase unusual reactions, positive and negative ion mass spectrometry of natural products using various ionization techniques, including DART, effects of metal cationization, LC/MS and MS/MS applications and quantitative analysis of drugs and metabolites. He authored more than 150 research publications. He was a member of the editorial board of *Journal of Mass Spectrometry* during 1995–2007. He is a fellow of the National Academy of Sciences, Allahabad, India. At present, he lives in Kochi.

List of Abbreviations and Units

°C	degree celsius
µg	microgram
µL	microliter
APCI	atmospheric pressure chemical ionization
API	atmospheric pressure ionization
BPC	base peak chromatogram
CAD	collision activated dissociation
CE	capillary electrophoresis
CE	collision energy
CID	collision induced dissociation
CXP	cell exit potential
Da	dalton
DAD	diode array detection
DP	declustering potential
EP	entrance potential
ESI	electrospray ionization
FDA	Food and Drug Administration
FIA	flow injection analysis
g	gram
GC-MS	gas chromatography-mass spectrometry
GS1	nebulizer gas
GS2	heater gas
h	hour
HPLC	high performance liquid chromatography
ICH	International Conference on Harmonization
IS	internal standard
IT	ion trap
kPa	kilopascal
L	liter
LC	liquid chromatography
LOD	limit of detection

LOQ	limit of quantification
LTQ	linear trap quadrupole
m/z	mass-to-charge ratio
mg	milligram
min	minute
mL	milliliter
mM	millimolar
MRM	multiple reaction monitoring
MS	mass spectrometry
ms	millisecond
MS/MS	tandem mass spectrometry
ng	nanogram
NMPB	National Medicinal Plants Board
NMR	nuclear magnetic resonance
PCA	principal component analysis
PDA	photodiode array detection
psi	pressure per square inch
QqQ$_{LIT}$	hybrid linear ion trap triple quadrupole
QTOF	quadrupole time of flight
r^2	correlation coefficient
RDA	retro-Diels-Alder
RSD	relative standard deviation
S/N	signal-to-noise ratio
SD	standard deviation
t_R	retention time
UPLC	ultra performance liquid chromatography
UV	ultraviolet
WHO	World Health Organization
XIC/EIC	extracted ion chromatogram

Rauvolfia Ethno- And Phytopharma-cological Review

1

1.1 INTRODUCTION

India, the land of Ayurveda (science of life), has a rich heritage of traditional medicine. The origin of Ayurveda dates back to the Vedic era. It is the ancient Indian system of healthcare and longevity (Varier, 2016; Dev, 1999). It is remarkable that Ayurveda survived from antiquity to the present day. The other two codified Indian Systems of Medicine (ISM) are Unani introduced by Arabs and Persians in the 8th century AD (Ahmad, 2008) and Siddha originated in the southern parts of India in the pre Vedic period (Zysk, 2008). Since time immemorial plants have been the principal raw materials for traditional medicines worldwide. Plants are also the important ingredients in ISM. Ayurveda uses about 1,200 species of plants, while Siddha and Unani systems use 900 and 700 species, respectively (Chatterjee and Pakrashi, 1991; Kannaiyan, 2008). Indian folk medicine (tribal or indigenous medicine) uses more than 8,000 species of plants in their healthcare systems (Government of India, 2000; Sen and Chakraborty, 2015). For centuries, indigenous people have been using medicinal plants for their healthcare. Even now, about 70–80% of the world's population rely on traditional, largely herbal, medicine to meet their primary healthcare needs (Farnsworth and Soejarto, 1991).

Medicinal plants have been used from ancient times by various ethnic groups within their traditional knowledge systems (Kumar et al., 2019).

The family Apocynaceae has about 250 genera and 2,000 species of tropical trees, shrubs and vines (Ng, 2006; Wiart, 2006). *Rauvolfia* is an important genus of the family Apocynaceae, its discovery dates back to the 16th century, and about 130–140 species are known till now (Vakil, 1955; Kunakh, 1996). In 1703, the French botanist Charles Plumier first coined the name *Rauvolfia* to commemorate the 16th-century German physician, botanist and traveler Leonhard Rauwolf (1535–1596) using the Latin version of Rauwolf's name (Plumier, 1703). Linnaeus used this spelling of *Rauvolfia* in his *Species Plantarum* (Linnaeus, 1753).

The pantropical genus *Rauvolfia* (devil's pepper) is a genus of evergreen trees and shrubs found in the tropical and subtropical regions of the world, including Central and South America, Africa, India, Sri Lanka, Burma, China, Malaysia and Java and Japan. The largest number of species is found in Africa and South America. There are mainly six species of *Rauvolfia* in India, namely, *R. serpentina* (L.) Benth. ex Kurz, *R. verticillata* (Lour.) Baill, *R. hookeri* Srinivas & Chithra, *R. micrantha* Hook f., *R. tetraphylla* Linn and *R. vomitoria* Afzel (Figure 1.1). *R. hookeri* Srinivas & Chithra and *R. micrantha* Hook f. are endemic to the Western Ghats of South India (Ahmedullah and Nayar, 1986; Bhattacharjee, 1998). *R. tetraphylla* Linn is a native of West Indies, but naturalized in India more than a century ago. *R. vomitoria* Afzel is an African species introduced in India.

India's wonder drug plant *R. serpentina* (L.) Benth. ex Kurz. – the Indian snakeroot, black snakeroot, serpentine wood or devil's pepper – is locally known as Sarpagandha, Nakuli, Chandrika, insanity herb or "pagal-ki-booti." It is an important medicinal plant native to the Indian subcontinent and

FIGURE 1.1   Different species of *Rauvolfia* found in India.

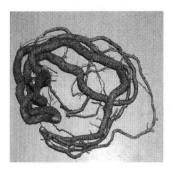

FIGURE 1.2   Root of *Rauvolfia*. (Reproduced from Muller, 2015 with permission from www.homeremediess.com)

Southeast Asian countries. *R. serpentina* has an ancient history (Monachino, 1954). The plant identified centuries later by the name of *R. serpentina* is mentioned by the name of Sarpagandha in the ancient Ayurvedic texts Charaka Samhita and Sushruta Samhita (Sastri, 2006; Somers, 1958; Bhishagratna, 1911). It has been in use for millennia in India for the treatment of stings and bites of insects and poisonous reptiles and mental illness. It is named snakeroot presumably because of the shape of the root and sarpagandha because it repels snakes (Figure 1.2).

   R. serpentina, Indian snakeroot, is an evergreen, erect glabrous perennial shrub having tuberous root with pale brown cork and growing up to a height of 60 cm and has tap root that reaches a length between 30 and 50 cm and a diameter between 1.2 and 2.5 cm. It is native to India and is widely distributed in India from the Himalayan region to the Eastern and Western Ghats and Andamans up to an altitude of 1,200 m. *R. verticillata*, syn. *R. densiflora*, dense flowered snakeroot, is an evergreen shrub growing up to 3–6 m tall with milky juice. It is distributed in the Indian subcontinent (North Eastern Hills, Eastern and Western Ghats), Southern China, Myanmar, Thailand, Malaysia, Indonesia, Laos, Cambodia, Vietnam and the Philippines. *R. micrantha* or small flowered snakeroot or Malabar *Rauvolfia* is a rare species endemic to southern Western Ghats found up to an altitude of 600 m. It is a perennial woody shrub growing up to 1.5 m. *R. micrantha* is used as a substitute for *R. serpentina* in commercial supplies (Youngken, 1954). The roots are a rich source of antihypertensive tranquilizer alkaloids and used as a substitute for *R. serpentina* to treat nervous disorders, especially in the state of Kerala (Sahu, 1979). *R. hookeri*, syn. *R. beddomei*, is a closely allied species, also endemic to the evergreen southern region of Western Ghats (Sahu, 1979; Ahmedullah and Nayar, 1986). It is a large, dichotomously branched shrub, 1.5–2 m in height and found up to an altitude of 700 m. *R. tetraphylla*, syn. *Rauvolfia canescens* L.;

Rauvolfia heterophylla Willd. ex Roem. & Schult, American serpent wood, wild snake root, be still tree, four leaved devil's pepper, Barachandrika, is a much branched woody shrub growing up to a height of 2 m. It is introduced from West Indies in 1845 (Woodson et al., 1957) and now naturalized in western and eastern peninsular India. It is also known as Garden Rauvolfia and is cultivated as an ornamental and medicinal plant (Farooqi and Sreeramu, 2004). It is commonly available in Bengal, Bihar, Odisha and Kerala. It is widely used in India as a substitute or adulterant of *R. serpentina*. *R. vomitoria*, serpent wood, swizzler stick, poison devil's pepper, African snakeroot, is a glabrous, erect medium sized tree growing up to 5–8 m tall at sea level and up to 1,600 m. The specific epithet "*vomitoria*" refers to the purgative and emetic properties of the bark. It is widely distributed in tropical Africa. In India, it is available as an introduced species and as an immigrant, got naturalized.

1.2 TRADITIONAL USES AND MEDICINAL PROPERTIES OF *RAUVOLFIA*

The plants of the genus *Rauvolfia* are important in Folklore medicine as well as modern medicine. *Rauvolfia serpentina* has been used in Indian folk medicine for thousands of years to treat a wide variety of maladies – including insomnia, hypertension, insanity, epilepsy, intestinal disorders, cardiac and liver diseases, hysteria, constipation, schizophrenia – and used as an anthelmintic, a tranquilizer and an antidote against the bites of snakes and other venomous reptiles (Mukerji, 1956). It is also used as a uterine stimulant, febrifuge and cure for insanity (Chatterjee and Pakrashi, 1995; Mukerji, 1976; Anonymous, 1999; Wilkins and Judson, 1953). The medicinal importance of the roots, leaves and juice of the plant has been recognized by practitioners of the indigenous system of medicine for a long time. Other species of the plant are also medicinally used in conventional Western medicine as well as in Ayurveda, Unani and folk medicine as the alkaloids in the plants reduce blood pressure, depress the activity of the central nervous system (CNS) and act as hypnotics, besides being an antidote for snake venom.

    R. serpentina has been in use since the Vedic period for the treatment of snake bites, insect stings, hypertension, insomnia, psychological disorders, gastrointestinal disorders, epilepsy, wounds, fever and schizophrenia (Chopra et al., 2009; Anonymous, 2006; Stöckigt, 1998). Different ethnic groups use *R. serpentina* to treat snake, insect and animal bites, mental illness, schizophrenia, hypertension, blood pressure, gastrointestinal diseases, circulatory disorders, pneumonia, fever, malaria, asthma, skin diseases, scabies, eye diseases, spleen diseases, AIDS, rheumatism, body pain, veterinary diseases, etc. (Dey and De, 2011). It is extensively used in the treatment of insanity and snake bite (Kokate et al., 2003). *R. verticillata* is used as a remedy for snake bite and hypertension (Sahu, 1979). Decoction of the roots reduces high blood pressure, and root powder with lime juice is applied for snake bites (Iqbal et al., 2013a). Tribal population use the leaf juice for diabetes, coughs, cold, peptic ulcers, stomachache and mouth infections, rheumatism, etc. (Iqbal et al., 2013a). *R. micrantha* is also used, especially in the state of Kerala, as a substitute for *R. serpentina* to treat a variety of nervous disorders (Sahu, 1979). *R. hookeri* is also used as a substitute for *R. serpentina*. *R. tetraphylla* exhibits significant activities and is widely used by South Indian tribes to stimulate uterine contraction, to alleviate stomachache and muscular pain, to cure cough and cold and skin diseases, and to treat mental disorders and high blood pressure (Iqbal et al., 2013b). According to Ayurveda, its root is bitter, acrid, pungent and anthelmintic and the extract preparations are used as antihypertensive, sedative and cure for various CNS disorders associated with psychosis, schizophrenia, insanity, insomnia and epilepsy. The plant is known in folk medicines of Sri Lanka as a treatment for snake poisoning and for external application in skin ailments (Arambewela and Madawela, 2001). *R. vomitoria* is traditionally used in African countries for hypertension, mental disorders, psychiatric management, malaria, typhoid and jaundice (Aquaisua et al., 2017; Fapojuwomi and Asinwa, 2013). *R. vomitoria* is often described as a plant with all parts poisonous, the roots and leaves with emetic and cathartic properties, and the bark used to treat fever and indigestion (Li et al., 1995). Traditionally, it has been used for its medicinal properties by African indigenous tribes (Cunningham, 1993). In traditional herbal medicine, the root was brewed as a tea and used in humans to treat hypertension, insanity, snake bite and cholera. Nigerian traditional healers use it to treat psychiatric patients. The bark has emetic and purgative properties. The root extracts have abortifacient properties (Omotayo and Borokini, 2012). The alkaloids in *R. vomitoria* are antihypertensive and sedative. The leaves, root and root and stem barks are extensively used for convulsions, fever, insomnia, mental disorders, high blood pressure, diabetes, snake bites, and skin infections and as anthelmintic. In India, *R. vomitoria* is used as a substitute for *R. serpentina*.

R. serpentina, *R. tetraphylla* and *R. vomitoria* have commercial importance, and presently, they are used as the best source of therapeutically active antihypertensive alkaloids (Rohela et al., 2016). The three main medicinal uses of *Rauvolfia* species are as follows: (1) raw material for the extraction of isolated alkaloids, (2) preparation of extracts with a standardized alkaloid content and (3) production of powdered *Rauvolfia* roots. The main uses of the pharmaceutical derivatives are as antihypertensives and sedatives. There is extensive folk medicinal use of the root bark and roots, particularly for their aphrodisiac, emetic, purgative, antipsoric, dysenteric, sedative, abortive and insecticidal properties. Decoctions prepared from the leaves of *R. vomitoria* have a powerful emetic effect. For swellings, a stew made from chopped leaves and animal fat is applied. *R. serpentina* root preparations have been used for centuries in India for treating disorders of the CNS, intestinal disorders and snake bites, as an anthelmintic and for the stimulation of uterine contractions.

1.3 PHYTOCHEMICAL CONSTITUENTS

The phytocomponents in *Rauvolfia* species consist of alkaloids, iridoids, flavonoids, terpenes, sterols, sugars and fatty acids. The most important group of bioactive phytochemicals found are indole alkaloids that are mostly concentrated in the roots. Genus *Rauvolfia* is known to accumulate considerable amounts of alkaloids. For example, 122 alkaloids have been reported from *R. serpentina* (Pathania et al., 2013). The genus is distinguished by the presence of some important monoterpenoid indole alkaloids (MIAs) such as ajmaline, deserpidine, rescinnamine, reserpine, serpentine and yohimbine, which exhibit a diverse array of structures and biological activities. The three types of alkaloids present are weak basic indole alkaloids, alkaloids of intermediate basicity and strong anhydronium bases. Though all parts of the plant *R. serpentina* contain alkaloids, the root has the highest amount, about 85–90% of the total alkaloid content (Pathania et al., 2013). The yield of total alkaloids in *R. serpentina* ranges from about 0.8% to 1.3% of the dry weight of the plant (Woodson et al., 1957). The most important alkaloid in *R. serpentina* is reserpine, first isolated in 1952 (Müller et al., 1952). Besides MIAs, the other classes of compounds present are anhydronium bases, phytosterols, phenols and glycosides. Other species of the genus, including *R. verticillata*, *R. tetraphylla*, *R. vomitoria*, *R. micrantha* and *R. hookeri*, contain variable amounts of indole alkaloids and may be utilized as alternative sources for the bioactive molecules. Ethno- and phytopharmacological overviews on *R. serpentina* (Kumari et al., 2013), *R. verticillata* (Iqbal et al., 2013a) and *R. tetraphylla* (Iqbal et al., 2013b; Jakaria et al., 2016; Mahalakshmi et al., 2019) are available in the literature.

According to a review in 2001 on phytochemical studies of *Rauvolfia* species, about 145 indole alkaloids have been reported (Ganapaty et al., 2001). An extensive library of molecules reported from *R. serpentina* has recently been compiled (Pathania et al., 2013). An exhaustive resource of *R. serpentina* plant derived molecules has been prepared in the form of a database

(Pathania et al., 2015). About 147 plant derived molecules from *R. serpentina* comprising 122 alkaloids, seven iridoid glucosides, six phenols, four each of anhydronium bases and phytosterols, three glycosides and one fatty acid are reported. Eighty six phytomolecules were from roots/stem/leaves/root bark, but mostly from roots of *R. serpentina*. Fifty five molecules were exclusively from cell culture, including hairy root culture, and three were common to both the plant and cell culture. The plant part was not specified for six molecules (Pathania et al., 2013). In a more recent review, 224 MIAs, two other alkaloids and fourteen nonalkaloidal compounds have been listed from *Rauvolfia* species (Boğa et al., 2018). There are a few reviews on the phytochemistry of *R. tetraphylla* (Iqbal et al., 2013b; Jakaraia et al., 2016).

Quercetin, kaempferol and rutin were isolated from alcoholic extracts of *R. serpentina* (Gupta et al., 2015; Gupta and Gupta, 2015a, 2015b). Analysis of flavonoid chromatographic (high performance liquid chromatography (HPLC)) profiles of chloroform extracts of acid hydrolyzed methanol extracts of leaves of *R. serpentina*, *R. verticillata* (*R. densiflora*), *R. tetraphylla*, *R. hookeri* (*R. beddomei*) and *R. micrantha* from the southern Western Ghats of India revealed some interesting results (Nair et al., 2013). Quercetin was detected in all five samples, myricetin and cyanidin only in *R. serpentina* and *R. tetraphylla*, kaempferol only in *R. hookeri*, luteolin only in *R. verticillata*, apigenin only in *R. hookeri* and *R. micrantha*, and delphinidin only in *R. serpentina* (Nair et al., 2013).

Indobine, a new alkaloid, was isolated and identified from *R. serpentina* (Siddiqui et al., 1987). Five new indole alkaloids, namely, N(b)-methylajmaline, N(b)-methylisoajmaline, 3-hydroxysarpagine, yohimbinic acid and isorauhimbinic acid, were isolated from the dried roots of *R. serpentina* (Itoh et al., 2005). 21-*O*-Methylisoajmaline, a new ajmaline type alkaloid, was isolated from the roots of *R. serpentina* together with twenty one known compounds, a mixture of β-sitosterol and stigmasterol, reserpinine, tetrahydroalstonine, reserpine, venoterpine, yohimbine, 6'-*O*-(3,4,5-trimethoxybenzoyl) glomeratose A, isoajmaline, 3-epi-α-yohimbine, methyl 3,4,5-trimethoxy-trans-cinnamate, a mixture of β-sitosterol 3-*O*-β-D-glucopyranoside and stigmasterol 3-*O*-β-D-glucopyranoside, rescidine, 7-deoxyloganic acid, ajmaline, suaveoline, (+)-tetraphyllicine, loganic acid, 3-hydroxysarpagine and sarpagine (Rukachaisirikul et al., 2017). Water extract of the powdered seeds of *R. serpentina* contained polysaccharides, which on methylation and hydrolysis with sulfuric acid provided three each of the methylated derivatives of glucose and mannose (Pathak et al., 2012).

Reserpine, ajmaline, vellosimine, spegatrine, verticillatine and dispegatrine are the main alkaloids from the roots of *R. verticillata* (Lin et al., 1985). Rescinnamine, isoreserpinine, reserpiline, reserpinine, sarpagine and densiflorine are the other alkaloids reported from *R. verticillata* (Iqbal et al., 2013a).

From the chloroform extract of *R. verticillata*, three indole alkaloids and one acridone alkaloid were isolated and identified as ajmalicine B, sandwicine, raunescine and 7-hydroxynoracronycine (Hong et al., 2012a). The acridone alkaloid is a new type of compound isolated from *Rauvolfia* genus for the first time (Hong et al., 2012b). Five new hexacyclic MIAs, namely, rauvovertine A, 17-epi-rauvovertine A, rauvovertine B, 17-epi-rauvovertine B and rauvovertine C, have been isolated from *R. verticillata* (Gao et al., 2015a).

The roots of *R. tetraphylla* yield the drug deserpidine, which is antihypertensive and tranquilizer. Twenty two indole alkaloids were reported from *R. tetraphylla* (Anonymous, 2005; Iqbal et al., 2013b; Jakaria et al., 2016). Ten indole alkaloids, namely, ajmaline, yohimbine, α-yohimbine, isoreserpine, corynanthine, deserpidine, reserpiline, isoreserpiline, arcine and a new alkaloid lankanescine, were isolated and identified from *Rauvolfia canescens (R. tetraphylla)* from Southern Sri Lanka (Arambewela and Madawela, 2001). Five new indole alkaloids rauvotetraphyllines A–E, together with eight known analogues, namely, alstonine, nortetraphyllicine, peraksine, sarpagine, 3-hydroxysarpagine, dihydroperaksine, hydroxydihydroperaksine and raucaffricine, were isolated from the aerial parts of *R. tetraphylla* (Gao et al., 2012). Five new indole alkaloids, namely, rauvotetraphyllines F-H, 17-epi-rauvotetraphylline F and 21-epi-rauvotetraphylline H, were isolated from the aerial parts of *R. tetraphylla* (Gao et al., 2015b). A new labdane diterpene 3β-hydroxy-labda-8(17), 13(14)-dien-12(15)-olide was isolated from air dried roots of *R. tetraphylla* (Brahmachari et al., 2011).

Seventy two alkaloids classified into nineteen types occur in *R. vomitoria*. Iwu and Court isolated forty three indole alkaloids from the stem bark of *R. vomitoria*; thirty nine were identified and two partially characterized. Heteroyohimbines (especially reserpiline) and N (a)-demethyldihydroindoles were the major alkaloids (Iwu and Court, 1982). In an earlier work, twenty eight alkaloids from *R. vomitoria* roots were isolated, twenty two were identified and six were in traces. Aricine, carapanaubine, carapanaubine-N-oxide, rauvoxine and suaveoline were isolated for the first time from *R. vomitoria* roots (Iwu and Court, 1977). From the stem bark of Nigerian *R. vomitoria*, twenty-two indole alkaloids were isolated and twenty characterized, and the alkaloids included E-seco heteroyohimbine, sarpagan, dihydroindole, yohimbine and heteroyohimbine (Sabri and Court, 1978). N-Methyl ajmaline, ajmalidine, ajmalinine and neoajmaline, along with ajmaline, reserpine and tetraphyllicine, were reported for the first time in *R. vomitoria* (Malik and Siddiqui, 1979).

Leaves of *R. vomitoria* contained three indole alkaloids (aricine, tetrahydroalstonine and isoreserpiline) and three α-oxindole-type isomers (carapanaubine, rauvoxine and rauvoxinine), but no reserpine was found (Patel et al., 1964). Two new dihydroindole alkaloids (sandwicine and isosandwicine) were isolated from *R. vomitoria* (Ronchetti et al., 1971). Two new indole alkaloids, 3-epi-rescinnamine and 3,4-dimethoxybenzoyl-reserpic acid methyl ester, have been isolated from the root bark of *R. vomitoria* (Lovati et al., 1996). Nineteen indole alkaloids of E-*seco* indole, sarpagan, picrinine, akuammiline, heteroyohimbine, oxindole, yohimbine and indolenine types were isolated from Ghanaian *R. vomitoria* leaves (Amer and Court, 1980). Two unusual nor-monoterpenoid indole alkaloids rauvomine A and rauvomine B, together with two known compounds peraksine and alstoyunine A, were isolated from the aerial parts of *Rauvolfia vomitoria* (Zeng et al., 2017). Reserpine content of *R. vomitoria* from Africa is more than twice that of *R. serpentina* found in India. As *R. serpentina* is considered an endangered herb, it is being replaced with the nonendangered *R. vomitoria* in synergistic herbal formulas.

R. serpentina exhibited the highest total phenolic content, while *R. tetraphylla* had the highest flavonoid content among the five species. *R. serpentina* showed the highest 2,2-diphenyl-1-picrylhydrazyl (DPPH) radical scavenging activity and also the highest pigment composition and vitamin E content, while *R. verticillata* (*R. densiflora*) showed the highest level of vitamin C content and metal-chelating activity among the five species. *R. tetraphylla* revealed the highest concentration of β-carotene. Lycopene was found in very low amounts while comparing with other nutrient compositions, and the maximum amount was in *R. tetraphylla* and the least amount was in *R. beddomei* (Nair et al., 2012).

1.4 PHARMACOLOGICAL ACTIVITY

As the main constituents of *Rauvolfia* are the monoterpene indole alkaloids, they contribute to the major bioactivities of *Rauvolfia*. The pharmacological and medicinal properties and structural significance of bioactivities exerted by the alkaloids of Apocynaceae are discussed in a recent review (Dey et al., 2017). Ethno- and phytopharmacological overviews on *R. serpentina* (Kumari et al., 2013), *R. verticillata* (Iqbal et al., 2013a) and *R. tetraphylla* (Iqbal et al., 2013b; Jakaria et al., 2016; Mahalakshmi et al., 2019) are available in the literature. Five of the *Rauvolfia* alkaloids, namely, reserpine, reserpinine, deserpidine, ajmalicine and ajmaline, are used in medicine. The pharmacological activities of constituents of *Rauvolfia* are given in Tables 1.1 and 1.2. Reserpine was a mainstay in the management of hypertension before the advent of the current pharmaceutical options for hypertension (beta blockers, calcium channel blockers and angiotensin-converting inhibitors). Reserpine acts via the CNS to reduce sympathetic tone, increase parasympathetic activity and

TABLE 1.1   Pharmacological activities of *Rauvolfia* constituents

ALKALOID	TYPE	FUNCTION	REFERENCE
Ajmalicine	Indoline alkaloids	Antihypertensive, vasodilator, tranquilizer, androgenic receptor antagonist	Roberts and Wink 1998, Wink et al. 1998; Wink 2015
Ajmaline	Terpene indole alkaloid (TIA)	Antiarrhythmic, Na^+/K^+ channel inhibitor	Rolf et al. 2003, Friedrich et al. 2007; Wink 2015
Alstonine	Indole alkaloid	Antipsychotic, anticancer	Elisabetsky and Costa-Campos 2006; Linck et al. 2011; Denis et al. 2006
Deserpidine	Indole alkaloid	Tranquilizer, antihypertensive	Schneider et al. 1955
Rescinnamine	Weakly basic indole alkaloids	Antihypertensive, sedative	Cronheim et al. 1954; Klohs et al. 1954
Reserpiline	Indoline alkaloids of intermediate basicity	Antihypertensive, antipsychotic	Hariga 1959; Gupta et al. 2012
Reserpine	Indole alkaloid	Antipsychotic, Antihypertensive, monoamine transport blocker, tranquilizer, antidepressant, anticancer	Colpaert 1987; Mahata et al. 1996; Wink 2015
Serpentine	Basic anhydronium alkaloids	Antihypertensive, tranquilizer, antioxidant, possible anticancer	Sachdev et al. 1961; Dutta et al. 2011; Beljanski and Beljanski 1986
Tetraphyllicine		Myocardial excitation	Duncan and Nash 1970
Yohimbine	Indole alkaloid	Aphrodisiac agent, alpha adrenoreceptor inhibitor	Lambert et al. 1978; Hai-Bo et al. 2013

TABLE 1.2   Some of the major pharmacological activities of *Rauvolfia*

S. NO.	ACTIVITY	EXTRACT	REFERENCE
1	Anthelmintic	Ethanolic extract	Tekwu et al. 2017
2	Antibacterial	Methanolic extract of root	Negi et al. 2014; Suresh et al. 2008; Patel et al. 2013
3	Anticancer activity (ovarian)	Ethanolic extract of root	Yu et al. 2013
4	Antidiabetic	Methanolic extract of root	Azmi and Qureshi 2016
5	Antidiarrheal	Methanolic extract of leaf	Ezeigbo et al. 2012
6	Antifungal	Ethanolic extract of leaf	Suresh et al. 2008
7	Antihypertension	Aqueous extract of leaf	Lobay 2015
8	Anti-inflammatory	Aqueous extract of root	Rao et al. 2012
9	Antimicrobial activity	Ethanolic extract of root	Deshmukh et al. 2012; Owk and Lagudu 2016
10	Antioxidant	Methanolic extract of leaf	Nair et al. 2012
11	Antiproliferative activity	Ethanolic extract of leaf	Deshmukh et al. 2012
12	Antivenom	Ethanolic extract of whole plant	Rajashree et al. 2013
13	Cardioprotective	Ethanolic extract of leaf	Nandhini and Bai 2015
14	Cardiotonic	Aqueous extract of leaf	Thinakaran et al. 2009
15	Cytotoxic	Chloroform extract of leaf	Behera et al. 2016
16	Hepatoprotective	Aqueous ethanolic extract of rhizome	Gupta et al. 2006a
17	Hyperglycemic	Methanolic extract of root	Azmi and Qureshi 2013
18	Hypoglycemic	Methanolic extract of root	Thinakaran et al. 2009; Pathania et al. 2013
19	Hypolipidemic	Aqueous extract of root	Qureshi and Udani 2009; Pathania et al. 2013

help normalize blood pressure. Reserpine is the most commonly used alkaloid for treating mild to moderate essential hypertension. Reserpine produces a tranquilizing effect by the depletion of catecholamines in the brain.

Rauvolfia is mainly used as (1) hypotensive, (2) tranquillizer, (3) sedative, (4) stimulant to the central peripheral nervous system, (5) stimulant to uterine contraction and (6) anthelmintic bitter tonic and febrifuge. The root extract of the genus *Rauvolfia* is used as an important ingredient in drugs for the treatment of hypertension, high blood pressure, mental illness, snake bites and problems related to CNS. Besides having sedative, aphrodisiac and antispasmodic properties, the root extract also possesses hypoglycemic and hypolipidemic activities against animal models (Qureshi et al., 2009; Pathania et al., 2013). Methanolic extracts of *R. serpentina* roots improve the glycemic, antiatherogenic, coronary risk and cardioprotective indices in alloxan-induced diabetic mice, apparently due to the presence of polyphenolic compounds along with alkaloids (Azmi and Qureshi, 2012). It not only improves hyperglycemic but also hematinic status of alloxan-induced diabetic mice (Azmi and Qureshi, 2013). According to the computational docking studies carried out with *R. serpentina* alkaloids, serpentine, ajmalicine, yohimbine, reserpine and ajmaline, it was found that they possess substantial potential to activate insulin receptor (Ganugapati et al., 2012). Methanolic extracts of *R. serpentina* exhibited strong antibacterial activity against most of the tested human pathogenic bacteria. Significant antidiarrheal activity was noticed in an experiment involving castor oil-induced diarrhea in mice after the administration of methanol extract of *R. serpentina* leaves, thus supporting the traditional use for treating diarrhea (Ezeigbo et al., 2012). Aqueous ethanolic extract of the rhizome of *R. serpentina* was found to be a promising hepatoprotective agent (Gupta et al., 2006a).

There are several reports on the pharmacology of extracts of *R. serpentina* showing hypotensive effect (Lobay, 2015). It is suggested that the hypotensive effect of *Rauvolfia* alkaloids is mediated solely via CNS mechanisms (McQueen et al., 1954). Reserpine is working as tranquilizer and also lowers the blood pressure. Serpentine, a weak hypotensive agent, and sarpagine have only a fleeting effect on blood pressure. The alkaloid yohimbine is hypotensive, a depressant of cardiovascular system and also a hypnotic. Through molecular docking studies, it was found that *R. serpentina* has potential inhibitory molecules on phospholipase A2, acetylcholinesterase, L-aminoacid oxidase, serine protease and proteolase, thus substantiating the traditional use of the plant as an antidote to snake bites (Sreekumar et al., 2014). Ajmaline stimulates respiration and bowel movement and is also useful in the management of arrhythmic heart disorders (Köppel et al., 1989). *R. serpentina* may have a promising role as a prophylactic drug in stroke-induced experimental dementia due to its neuroprotective effect (Kanyal, 2016).

Pharmacological activity of *R. verticillata* has been reviewed (Iqbal et al., 2013a). The crude extracts of *R. densiflora* (*R. verticillata*) showed sedative properties as the leaf extract significantly reduced dips and rearing using the rat hole board technique (Weerakoon et al., 1998). Seeds and roots of *R. verticillata* have been used as anti-snake venom and for curing skin diseases (Kumar et al., 2019).

Dry fruit extracts of *R. tetraphylla* were assayed against eight bacterial species and found to possess potential broad-spectrum antimicrobial activity (Alagesaboopathi, 2009; Abubacker and Vasantha, 2011). Antibacterial activity of ethanol extract from *R. tetraphylla* was tested against bacterial species of *Escherichia coli, Streptococcus lactis, Enterobacter aerogenes, Alcaligenes faecalis, Pseudomonas aeruginosa* and *Proteus vulgaris*. The extracts showed the maximum activity against *E. coli, E. aerogenes* and *A. faecalis* (Suresh et al., 2008). Root bark of *R. tetraphylla* showed good *in vitro* antibacterial activity and *in vivo* anti-inflammatory activity in rats (Rao et al., 2012). Methanol and chloroform extracts of *R. tetraphylla* showed good antimicrobial activity against most gram-positive and gram-negative bacteria (Patel et al., 2013). The new labdane diterpene, 3β-hydroxy-labda-8(17), 13(14)-dien-12(15)-olide, isolated from air-dried roots of *R. tetraphylla* showed significant antitumor and anticancer activities when tested against human cancer cell lines (Brahmachari et al., 2011). *R. tetraphylla* alkaloidal chloroform fraction at pH 9 showed high antipsychotic activity against dopaminergic and serotonergic receptors *in vitro* and amphetamine-induced hyperactive mouse model *in vivo* (Gupta et al., 2012). Activity-guided isolation of chloroform fraction afforded six indole alkaloids: 10-methoxytetrahydroalstonine, isoreserpiline, an isomeric mixture of 11-demethoxyreserpiline and 10-demethoxyreserpiline, α-yohimbine and reserpiline. Leaf and fruit extracts of *R. tetraphylla* showed good antioxidant and cytotoxic activities (Vinay et al., 2016; Behera et al., 2016). Antihypertensive effect of methanolic extract of *R. tetraphylla* was tested on deoxycorticosterone acetate (DOCA)-salt-induced hypertensive rats, and it was found that oral administration of the plant extract resulted in a remarkable reduction in systolic blood pressure (Gadvi et al., 2018). It was also demonstrated that leaves of *R. tetraphylla* had a significant anticonvulsant activity (Singh et al., 2019).

The antihypertensive effect of *R. vomitoria* has been reviewed (Stansbury et al., 2012). The crude aqueous root bark extract of *R. vomitoria* dose dependently decreased pain perception (analgesic effect), and it appears to have a high potential as an antipsychotic agent (Bisong et al., 2011). The effect of aqueous extract of *R. vomitoria* root bark on the histology of the cerebellum and neurobehavior pattern of adult male Wistar rats was investigated, and a reduction in body weight, mild distortions of the cerebellum and reduction in locomotion and exploratory behaviors were found. It is recommended, therefore, that, though used effectively to treat psychotic disorders, it may cause different adverse effects in individuals (Eluwa et al., 2008).

The root extract of *R. vomitoria* also shows antidiabetic activity (Campbell-Tofte, 2009). It was reported that the aqueous extract of roots without bark of *R. vomitoria* is nontoxic when taken orally and has high hypoglycemic and antihyperglycemic activities justifying the use of this plant in traditional medicine for the treatment of diabetic hyperglycemia (N'doua et al., 2016). The ethanolic root extract of *R. vomitoria* has potent antitumor activity and in combination significantly enhances the effect of carboplatin against ovarian cancer (Yu et al., 2013). A bioactive carboline alkaloid, alstonine, present in the root and leaves of *R. vomitoria* has anticancer activity (Denis et al., 2006). During an *in vitro* assessment of anthelmintic activities, it was found that the ethanol crude extracts of *R. vomitoria* stem bark and root possess moderate antischistosomal properties against two life stages of *S. mansoni*: cercariae and adult worms (Tekwu et al., 2017). The aqueous ethanolic and alkaloid extracts of the roots of *R. vomitoria* exhibited moderate to high antimycobacterial activity against *Mycobacterium madagascariense* and *M. indicus pranii* (Erasto et al., 2011). Ethanolic extract of the stem bark of *R. vomitoria* exhibited *in vitro* antiplasmodial activity (Zirihi et al., 2009). Aqueous and methanol extracts of the leaves of *R. vomitoria* also exhibited antiplasmodial activity (Cynthia et al., 2018). Toxicity studies of leaf and root extracts of *R. vomitoria* in Wistar rats revealed no significant toxicity problems to any organs in the concentrations employed for the study (Ebuehi et al., 2018). The methanol extract of *R. vomitoria* exhibited significant anthelmintic and antioxidant activities (Adu et al., 2015). Aqueous leaf extract of *R. vomitoria* demonstrated potential anticonvulsant properties (Olatokunboh et al., 2009). The ethanol extract of *R. vomitoria* suppressed the growth of *P. berghei* NK65, and it had antihyperglycemic and hypolipidemic effects on mice infected with *P. berghei* NK65 (Momoh et al., 2014). Antipancreatic cancer activity of *R. vomitoria* extract was demonstrated during *in vitro* and *in vivo* experiments (Yu and Chen, 2014; Dong et al., 2018).

1.5 PHYTOCHEMICAL ANALYSIS

1.5.1 HPTLC, HPLC, GC–MS, CE–MS and LC–MS Analysis

Several analytical techniques such as high-performance thin-layer chromatography (HPTLC), gas chromatography (GC), gas chromatography mass spectrometry (GC–MS), capillary electrophoresis mass spectrometry (CE–MS), high-performance liquid chromatography (HPLC), and high-performance liquid chromatography mass spectrometry (HPLC–MS) have been used for the analysis of *Rauvolfia* extracts. GC is not convenient, easy or reliable for the analysis of *Rauvolfia* alkaloids. Reserpine was hydrolyzed and 3, 4, 5-trimethoxybenzoic acid generated esterified by diazomethane and the resulting methyl ester was quantified by GC (Settimj et al., 1976). Trimethyl silylation followed by separation on a 2-m column at 270°C is also reported (Forni, 1979). An HPTLC method was developed and validated for the quantification of reserpine in *R. serpentina* and its allied preparations (Lohani et al., 2011). A validated densitometric HPTLC method was developed for the simultaneous quantification of reserpine and ajmalicine in *R. serpentina* and *R. tetraphylla* (Pandey et al., 2016). The normal-phase separation and reverse-phase separation of indole alkaloids were achieved by HPTLC and HPLC, respectively (Gupta et al., 2006b). HPTLC has been used for the determination of reserpine in *R. serpentina* homoeopathic mother tinctures (Dwivedi et al., 2017). HPTLC estimation of yohimbine in stem, leaf and seed of *R. tetraphylla* showed that the leaf contained the maximum amount of yohimbine (Kumar et al., 2011). HPLC and HPTLC methods were developed for the determination of indole alkaloids from plant cell callus cultures of *R. serpentina* and hairy roots of *R. serpentina* and *R. vomitoria* (Elyushnichenko et al., 1995). A simple isocratic HPLC method has been developed for the simultaneous quantitation of three antipsychotic indole alkaloids (α-yohimbine, isoreserpiline and 10-methoxy tetrahydroalstonine) in the leaf of *R. tetraphylla* (Verma et al., 2012). In an HPLC study on flavonoids as chemotaxonomic markers in endemic/endangered species of *Rauvolfia* (*R. serpentina, R. verticillata, R. hookeri, R. micrantha, R. tetraphylla* and *R. vomitoria*) from the southern Western Ghats of India, it was found that the flavonol quercetin was detected in all five samples, myricetin and cyanidin only in *R. serpentina* and *R. tetraphylla*, kaempferol only in *R. hookeri*, luteolin only in *R. verticillata*, apigenin only in *R. hookeri* and *R. micrantha*, and delphinidin only in *R. serpentina* (Nair et al., 2013).

Quantitation of yohimbine in biological fluids was reported using HPLC with amperometric detection (Diquet et al., 1984). Identification and determination of yohimbine using HPLC with UV detection was also reported (Obreshkova and Tsvetkova, 2016). An HPLC method was developed for the determination of reserpine in *R. vomitoria* and *R. serpentina* (Zhang et al., 2007). A novel method having good separation and low detection limit has been reported combining HPLC with online post-column electrochemical derivatization and fluorescence detection for the determination of reserpine in mouse serum (Chen et al., 2018). Chloroform extracts of *R. verticillata* were analyzed using HPLC-UV and GC-MS for monitoring the indole alkaloids sarpagine, yohimbine, ajmaline, ajmalicine and reserpine. Thirty-nine volatile compounds, including volatile oils, steroids and terpenes, were identified using GC-MS (Hong et al., 2013).

Several methods are available for the analysis of reserpine in tablets (Abdine et al., 1978), plants (Srivastava et al., 2006), urine (Li et al., 2011) and plasma (Anderson et al., 1997; Iqbal et al., 2013c). Reserpine in tablets was determined by a rapid and sensitive RP-HPLC method (Al-Akraa and Kabaweh, 2015). HPLC was utilized to determine the reserpine content of roots of various genotypes across different age groups of *R. serpentina*, and it was found that the reserpine content in the roots of *R. serpentina* had positive variation with the age of the plant (Koul et al., 2017). The chemical patterns of different solvent extracts of *R. hookeri* were compared using HPLC and HPTLC profiling. Among the different solvents used for extraction, chloroform showed the maximum yield of reserpine (i.e., 0.06 µg/10 µL) and the maximum numbers of compounds in root extracts (Kurian et al., 2017). A new rapid UPLC-UV method using C_{18} column has been developed for the simultaneous analysis of eight alkaloids (rauwolscine, ajmaline, yohimbine, corynanthine, ajmalicine, serpentine, serpentinine and reserpine) from the root samples of *R. serpentina* and *R. vomitoria* (Sagi et al., 2015). A CE-MS method was developed for the identification of alkaloids in *Rauvolfia* roots, and the method was applied for analyzing the alkaloids formed in *Rauvolfia in vitro* cultures (Stöckigt et al., 1998).

A combined HPLC-DAD-MS, HPLC-MS[n] and NMR spectroscopy has been used for the quality control of plant extracts and commercial blends sold as dietary supplements (Karioti et al., 2014). HPLC and MS methods are the most convenient and sensitive (Hong et al., 2013; El-Din et al., 2016). Using LC-QTOF MS, twenty-two components of *R. verticillata* were identified by comparing retention times and molecular weights with those of available standards and with reference data (Hong et al., 2010). Commercial samples containing *Rauvolfia* were also analyzed, revealing a wide variation in the content of alkaloids.

Using papaverine as an internal standard, reserpine, rescinnamine and yohimbine in human plasma were determined using UPLC-MS/MS method (Iqbal et al., 2013c). An HPLC-MS/MS method using C18 column was reported for the simultaneous determination of yohimbine, sildenafil, vardenafil and tadalafil in dietary supplements (Zhang et al., 2010). Using UPLC-QTOF-MS/MS, twenty-six alkaloids were characterized (Sagi et al., 2016). A UPLC-ESI-Q-TOF method for the rapid and reliable identification and quantification of major indole alkaloids in *Catharanthus roseus* was developed (Jeong and Lim, 2018). UPLC coupled with Ion Mobility QTOF MS was also used for profiling indole alkaloids (Sun et al., 2011). Yohimbine has been detected in food supplements using UPLC-ESI-MS (Ivanova et al., 2017).

An interesting recent report describes a semi-quantitative analysis of the various MIAs in fruits and other parts of *R. tetraphylla* using UPLC-ESI-MS and DESI-MS (desorption electrospray ionization-mass spectrometry) imaging (Kumara et al., 2019). The results indicate distinct localization patterns across and within different tissues. The roots contained most of the MIAs, while the stems contained only sarpagine, ajmaline, serpentine, ajmalicine, yohimbine and 18-hydroxy yohimbine. Similarly, the leaves had only serpentine, ajmalicine, reserpiline and yohimbine. Using UPLC-ESI-MS/MS data, sixteen MIAs were identified (Kumara et al., 2019).

1.5.2 Adulteration/Endangered Species

The annual requirement of dry *Rauvolfia* in India is 650t (Kunakh, 1996). Overexploitation led to the depletion of natural resources, and because of the high demand, adulteration is common. Poor method of conventional propagation and indiscriminative use of *R. serpentina* have led to the inclusion of this species in the list of endangered plants by IUCN (CITES Appendix II) and in the list of endangered plants of The National Medicinal Plants Board (NMBP), Govt. of India. Because of the depletion of natural *Rauvolfia* thickets, there were efforts to raise *Rauvolfia* as an agricultural crop using more productive plant varieties or through cell culture. The method of adulteration varies with the availability of the resources. For example, the roots of *R. serpentina* may be adulterated with the stem of the same plant or roots of the other species such as *R. hookeri*, *R. micrantha*, *R. verticillata*, *R. tetraphylla* and *R. vomitoria*. These species serve as substitutes as they resemble the genuine herbal drug with some differences in the qualitative and quantitative contents of phytochemicals. All the *Rauvolfia* species available in India are used for the production of sedatives and cardiovascular medicines.

1.5.3 Quality Control

The popularity of traditional herbal medicines is increasing because of the time-tested therapeutic efficacy and little or no side effects (Kamboj, 2000; WHO, 2004; Ekor, 2014). However, inconsistency in the bioactive phytoconstituents in the herbal drugs significantly affects their efficacy and undermines the practice of herbal medicines itself (Yadav, 2008; WHO, 2013). The medicinal property of a herb is due to the synergistic effects of its phytoconstituents. It is, therefore, of immense importance to maintain the delicate balance of the phytoconstituents to achieve the desired therapeutic effects. The characteristic chemical constituent profile of each plant has to be maintained for uniform herbal drug efficacy. This requires strict chemical quality control of herbs and finished products. This is achieved by taking advantage of modern analytical techniques such as HPTLC, HPLC, NMR, LC-MS and LC-MS/MS. The potent medicinal plants of *Rauvolfia* species are taxonomically and morphologically similar but contain varying amounts of bioactive constituents (Mabberley, 2017; Bindu et al., 2014; Boğa et al., 2018). Identification and authentication of *Rauvolfia* species is difficult due to more morphological similarity, and hence, a rapid analytical procedure is required for their timely and convenient identification. In a continuing program on the fingerprinting of medicinal plants, our laboratory is involved in chemical fingerprinting and chemometric analyses of important Indian medicinal plants. In this connection, we have undertaken qualitative and quantitative analyses of the phytochemicals of six *Rauvolfia* species, namely, *R. serpentina*, *R. verticillata*, *R. hookeri*, *R. micrantha*, *R. tetraphylla* and *R. vomitoria*.

Reserpine content in the roots of six *Rauvolfia* species (*R. serpentina*, *R. verticillata*, *R. hookeri*, *R. micrantha*, *R. tetraphylla* and *R. vomitoria*) was detected by HPLC coupled with electrospray ionization quadrupole time of flight tandem mass spectrometry (HPLC-ESI-QToF-MS/MS) and estimated by validated HPTLC method (Bindu et al., 2014). Of the six *Rauvolfia* species, reserpine content was highest in the exotic species *R. vomitoria* (689.5 μg/g, dry wt.), while of the five Indian species, the reserpine content was highest in *R. tetraphylla* (450.7 μg/g, dry wt.). In the most common Indian *Rauvolfia* species, *R. serpentina*, the reserpine content was comparatively low (254.8 μg/g, dry wt.). The endemic species *R. micrantha* possesses a significant quantity of reserpine (422.1 μg/g, dry wt.), making it a potential candidate for developing as a source of reserpine, replacing *R. serpentina* and *R. tetraphylla* that are endangered due to overexploitation. *R. hookeri* had 132.3 μg/g, dry wt., whereas it was not detected in *R. verticillata* (Bindu et al., 2014).

For the quick assessment of herbal raw materials, a rapid method was developed for fingerprinting of roots and leaves of six *Rauvolfia* species by direct analysis in real-time mass spectrometry (DART-MS). Seventeen bioactive MIAs were tentatively identified on the basis of their exact mass measurement from the intact plant parts. Principal component analysis could discriminate the six *Rauvolfia* species (Kumar et al., 2015). Orbitrap MS^n analysis was carried out to elucidate the fragmentation pathways of MIAs in *Rauvolfia* species (Kumar et al., 2016a). A UPLC-triple quadrupole-linear ion trap mass spectrometry (UPLC-QqQLIT-MS/MS) method in multiple reaction monitoring (MRM) mode was developed for the simultaneous determination of bioactive MIAs in the ethanolic extract of six *Rauvolfia* species and herbal formulations (Kumar et al., 2016c). An efficient and reliable liquid chromatography-QTOF tandem mass spectrometry (LC-QTOF-MS/MS) method was developed for the ethanolic root extract of *Rauvolfia* species to elucidate the fragmentation pathways for dereplication of bioactive MIAs (Kumar et al., 2016b). The details of some of these investigations are dealt with in this book.

Structural Characterization of Monoterpene Indole Alkaloids in Ethanolic Extracts of *Rauvolfia* Species by LC-QTOF-MS

2

2.1 INTRODUCTION

Rauvolfia species are characterized by the presence of bioactive monoterpene indole alkaloids (MIAs), such as reserpine, ajmalicine, ajmaline, serpentine, yohimbine, deserpidine and rescinnamine, responsible for their therapeutic effects. Because of depletion of the natural sources of *R. serpentina*, the other available species *R. verticillata*, *R. hookeri*, *R. micrantha*, *R. tetraphylla* and *R. vomitoria* are being used for herbal preparations. It is essential that a proper fingerprint of the phytochemicals in these species is made available so that the quality and efficacy of the herbal products can be ascertained and maintained at the desired level. Only a few reports are available for the identification and characterization of MIAs by liquid chromatography-tandem mass spectrometry (LC-MS/MS) in plant extracts (Hong et al., 2010; Kumar et al., 2015; Sun et al., 2011; Sagi et al., 2016; Chen et al., 2013; Liu et al., 2016; Uhlig et al., 2014). There is no report on the comparative and comprehensive phytochemical investigation of the roots of *Rauvolfia* species. It was, therefore, decided to develop a simple and specific high-performance liquid chromatography-electrospray ionization-quadrupole time-of-flight tandem mass spectrometry (HPLC-ESI-QTOF-MS/MS) method to establish fragmentation pathways for the identification and characterization of bioactive compounds of the ethanolic extracts obtained from the roots of *Rauvolfia* species.

2.2 METHODS USED FOR ANALYSIS

Herbal *Rauvolfia* preparations utilize the roots of the plants. The alkaloids are localized mainly in the roots. Hence, the roots of the plants are analyzed for the bioactive constituents. LC-QTOF-MS gives accurate mass and MS/MS data and thus helps to identify and characterize the components which come out of the HPLC system. Chemometric analysis such as principal component analysis (PCA) should provide discriminative outputs to distinguish the different *Rauvolfia* species (Ian, 2002).

2.2.1 Sample Collection

The roots of *R. serpentina*, *R. verticillata*, *R. hookeri*, *R. micrantha*, *R. tetraphylla* and *R. vomitoria* were collected from the plants grown under similar conditions in Jawaharlal Nehru Tropical Botanic Garden and Research Institute (JNTBGRI) campus, Kerala, India. The plant materials were collected in September 2012, and the voucher specimens (*R. serpentina* – 66451, *R. verticillata* – 66453, *R. hookeri* – 66449, *R. micrantha* – 66450, *R. tetraphylla* – 66452 and *R. vomitoria* – 66454) were deposited in the Herbarium of JNTBGRI.

2.2.2 Sample Preparation

The washed and dried roots were powdered, packed in airtight containers and stored at 20°C until analysis. The powdered roots (50 g each) were extracted with 250 mL of ethanol by sonication for 30 min at 30°C followed by keeping at room temperature (26°C–28°C) for 24 h. The extracts were filtered through filter paper (Whatman No. 1), and the residues were re-extracted four times using the same procedure with a fresh solvent. The combined filtrates were evaporated to dryness under reduced pressure (20–50 kPa) at 40°C using a rotary evaporator (Buchi Rotavapor-R2, Flawil, Switzerland). Stock solutions (1 mg/mL) of each of the dried plant extracts were prepared in methanol and filtered through a 0.22-μm polyvinylidene fluoride (PVDF) membrane into the HPLC auto sampler vial prior to LC-MS analysis.

2.3 LC-MS ANALYSIS OF PHYTOCHEMICALS

2.3.1 HPLC-ESI-QTOF-MS/MS Conditions

An Agilent 1200 HPLC system interfaced with Agilent 6520 hybrid quadrupole time-of-flight mass spectrometer (Agilent Technologies, USA) was used for the analysis of *Rauvolfia* root extracts. The 1200 HPLC system was equipped with a quaternary pump (G1311A), online vacuum degasser (G1322A), auto sampler (G1329A), thermostatted column compartment (G1316C) and diode array detector (G1315D).

2.3.1.1 Chromatographic Conditions

A Thermo Betasil C8 column (250 mm × 4.5 mm, 5 μ) operated at 25°C was used for HPLC analysis. The chromatographic separation was achieved using a gradient elution of 0.1% formic acid in water (A) and acetonitrile (B) as the mobile phases at a flow rate of 0.6 mL/min. The gradient had the following steps: 0–15 min, 25–27%; 15–18 min, 27–37%; 18–20 min, 37–42%; 20–22 min, 42–45%; 22–25 min, 45–48%; 25–30 min, 48–60%, and then returned to the initial conditions after 5 min. The sample injection volume was 1 μL.

2.3.1.2 Mass Spectrometric Conditions

The mass spectrometer was operated in positive ESI mode, and the spectra were recorded by scanning the mass range m/z 50–1,500 in both MS and MS/MS modes. Nitrogen was used as drying, nebulizing and collision gas. The drying gas flow rate was 12 L/min. The heated capillary temperature was set to 350°C and nebulizer pressure at 45 psi. The source parameters capillary voltage (VCap), fragmentor, skimmer and octopole voltages were set to 3,500, 175, 65 and 750 V, respectively. Product ion spectra of the standards (20 μg/mL in CH₃OH) were studied by direct infusion in ESI source using an infusion pump at a flow rate of 5 μL/min. For the MS/MS analysis, collision energy was set at 30, 35 and 40 eV. The accurate mass data of the molecular ions were processed using the Mass Hunter Workstation (version B 04.00) software (Agilent Technology, USA). HPLC-ESI-QTOF-MS data obtained from three repeats of all the samples were subjected to statistical analysis. PCA was performed on statistical software STATISTICA version 7.0 (StatSoft, Inc., USA).

2.3.2 Qualitative Analysis

When a series of compounds with similar structural features are analyzed, similar fragmentation patterns are observed even with different functional groups. It is therefore wise to start with reference standards. Plants of the *Rauvolfia* species are a rich source of alkaloids, and hence, as a prelude to structural characterization of MIAs from *Rauvolfia* species, five standards, namely, ajmalicine (4), ajmaline (19), yohimbine (28), serpentine (32) and reserpine (44), were selected as templates to generate diagnostic fragmentation pathways using LC-QTOF-MS. Being alkaloids, all of these compounds gave prominent protonated molecular ions. The [M+H]⁺ ions of these standards were selected and product ion spectra obtained (Figure 2.1).

MS/MS of the protonated yohimbine (*m/z* 355.2019) showed the initial neutral molecule losses of H_2O (*m/z* 337.1920) and CH₃OH (*m/z* 323.1560). retro-Diels-Alder (RDA) reactions and C-ring cleavages give rise to characteristic fragment ions. The fairly abundant peak at *m/z* 224.1285 corresponds to the loss of methyl indole (C_9H_9N), and that at *m/z* 212.1282 corresponds to the loss of $C_{10}H_9N$. The *m/z* values of these peaks clearly indicate the substitution on the terpene moiety. Similar C-ring cleavages with charge retention on the indole moiety lead to the most intense ion at *m/z* 144.0808 ($C_{10}H_{10}N^+$) and an accompanying ion at *m/z* 158.0959 ($C_{11}H_{12}N^+$). The *m/z* values of these peaks indicate the substitution on the indole moiety (Figure 2.2).

MS/MS of the protonated reserpine ion at *m/z* 609.2807 showed fragment ions corresponding to the losses of CH₃OH and trimethoxy benzoic acid at *m/z* 577.2492 and 397.2090, respectively. Elimination of 3,4,5-trimethoxy benzoic acid from *m/z* 577 or CH₃OH from *m/z* 397 leads to the ion at *m/z* 365.1835. C-ring cleavages gave rise to a characteristic fragment at *m/z* 448.1930 corresponding to the elimination of 3-methyl-6-methoxy indole ($C_{10}H_{11}NO$). An accompanying fragment was at *m/z* 436.1934, resulting from RDA followed by C, D-ring cleavages and elimination of $C_{11}H_{11}NO$. The shifts in the *m/z* values of these peaks indicate the substitution on the terpene moiety (Figure 2.2).

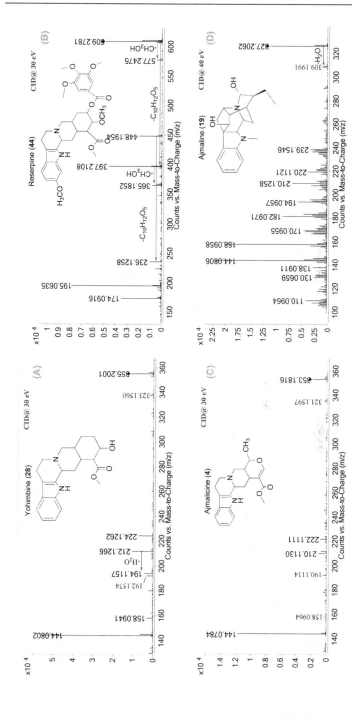

FIGURE 2.1 Q-TOF MS/MS spectra of yohimbine, reserpine, ajmalicine and ajmaline. (Reproduced from Kumar et al., 2016b with permission from Elsevier.)

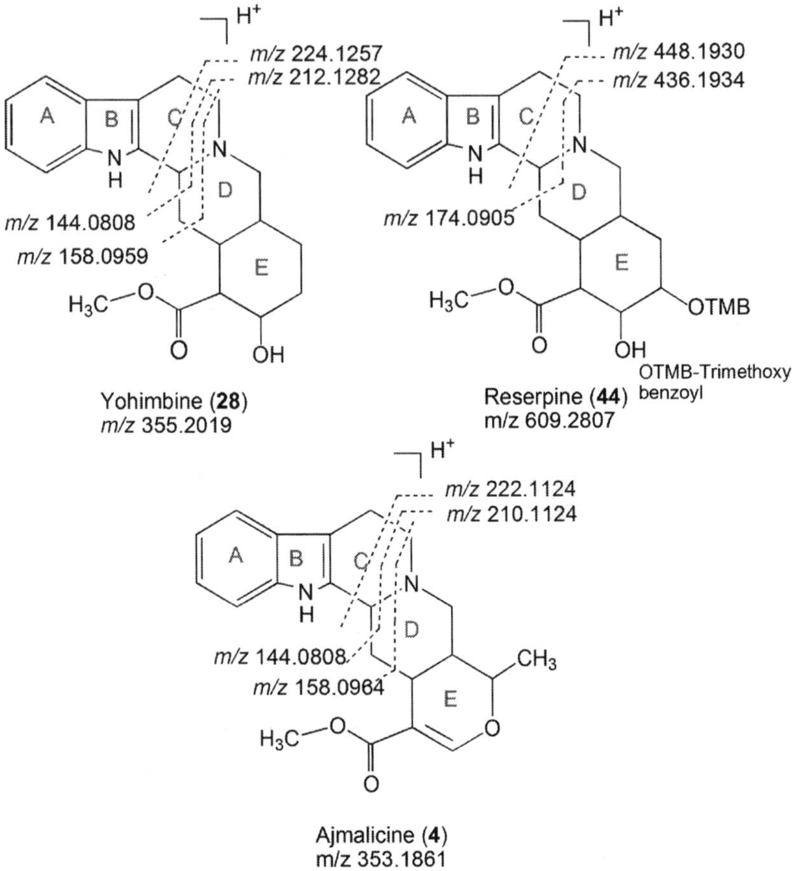

FIGURE 2.2 Ring cleavages of yohimbine (28), reserpine (44) and ajmalicine (4) in MS/MS.

C-ring cleavages with charge retention on the indole moiety lead to the fairly abundant peak at m/z 174.0905 ($C_{11}H_{12}NO^+$), which indicates the nature of substitution on the indole moiety (Figure 2.2). 3,4,5-Trimethoxy benzoic acid loss is also seen from the ion at m/z 448 resulting in the ion at m/z 236.1261. The most abundant ion corresponds to the 3,4,5-trimethoxy benzoyl ion (m/z 195.0640). Ajmalicine (m/z 353.1861) showed product ions at m/z 321.1597 (loss of CH_3OH), 222.1124 (corresponds to C-ring cleavage and loss of methyl indole), 210.111 (C, D-ring cleavages and loss of $C_{10}H_9N$), 144.0808 (C-ring cleavage with charge on the indole moiety) and 158.0964 (C-ring cleavage with charge on the indole moiety). These fragments indicate the substitution on the indole and terpene moieties (Figure 2.2). Successive losses of H_2O gave rise to the ions at m/z 309.1991 and 291.1881 in ajmaline (m/z 327.2067). The other fragments were at m/z 170.0974 (45), 158.0964 (100) and 144.0812 (82). C-ring cleavages produced these N-methyl indole derivative ions (Scheme 2.1).

The MS/MS spectrum of serpentine (m/z 349.1547) showed fragment ions at 317.1290 (loss of CH_3OH), 289.1314 (loss of CH_3OH and CO), 263.0796 (loss of $C_5H_{10}O$), 235.0826 (loss of CO from 263) and 207.0858 (loss of CO from 235). These fragmentations are possibly assisted by RDA reaction. Loss of substituents as neutral molecules and ring cleavages are the major MS/MS fragmentations of these templates.

2.3.3 Metabolic Profiling

The objective of metabolic profiling was to find out the phytochemical variations in the six selected *Rauvolfia* species, namely, *R. serpentina*, *R. verticillata*, *R. hookeri*, *R. micrantha*, *R. tetraphylla* and *R. vomitoria*. All of the six *Rauvolfia* species were cultivated in similar environmental conditions prior to harvesting the samples for the study. Ethanolic extracts of their roots were analyzed using a gradient mobile phase consisting of acetonitrile and 0.1% aqueous formic acid after optimizing parameters such as column type, column temperature, mobile phase, elution conditions, flow rate and MS conditions. Base peak chromatograms (BPCs) of ethanolic extracts of *R. serpentina*, *R. verticillata*, *R. hookeri*, *R. micrantha*, *R. tetraphylla* and *R. vomitoria* in positive-ion mode are shown in Figure 2.3. The peak numbers correspond to the compounds identified. Retention time (t_R), observed [M+H]$^+$, molecular formula, error (Δppm), major fragment ions and their relative abundance and distribution along with assignment are presented in Table 2.1. All of these compounds were identified based on their exact mass, molecular formula and fragmentation patterns of standards, and those available from the literature (Pathania et al., 2013, Boğa et al., 2018, Ramos et al., 2019). A total of 47 known/unknown MIAs were tentatively identified and characterized using the diagnostic fragmentation pathways of the templates. Yohimbine (28, t_R 10.91 min), reserpine (44, t_R 24.31 min), ajmalicine (4, t_R 5.52 min), ajmaline (19, t_R 8.91 min) and serpentine (32, t_R 15.40 min) were unambiguously identified and characterized by comparison with their authentic standards.

2.3.3.1 Reserpine Class of Compounds

Seventeen reserpine class compounds were tentatively identified and characterized by comparing the fragmentation pathways of their [M+H]$^+$ ions with those of the standards, yohimbine and reserpine eluting at 10.91 min and

24.31 min, respectively. Fragmentation pathways of reserpine class of compounds are shown in Figure 2.2. Compounds **1** (t_R 3.4 min) and **16** (t_R 7.81 min) were tentatively identified as yohimbinic acid and reserpic acid, respectively. The molecular masses of compound **1** (*m/z* 341.1867, 14 Da less than that of **28**) and compound **16** (*m/z* 401.2071, 208 Da less than that of reserpine) indicate a difference in substitution. Both compounds showed the loss of H_2O (*m/z* 323.1754 in compound **1** and 383.1965 in compound **16**), whereas compound **16** also showed the loss of CH_3OH (*m/z* 369.1809), but compound 1 did not. C-ring cleavages led to the ions at *m/z* 210.1119 (14 Da less than that in compound **28**) and *m/z* 158.0932 and 144.0794 (the same as in compound **28**) in compound **1**. C-ring cleavages in compound **16** resulted in peaks at *m/z* 240.1200 (208 Da less than that in compound **44**) and *m/z* 174.0907 (same as in compound **44**), indicating difference in substitution in the terpene part. Compound **16** did not show any loss of 3,4,5-trimethoxy benzoic acid, indicating the absence of 3,4,5-trimethoxy benzoyl group. Hence, compounds **1** and **16** are identified as yohimbinic acid and reserpic acid, respectively. Compounds **1** (t_R 3.4 min) and compound **11** (t_R 6.9 min) gave the same fragment ions with different relative abundance in their MS/MS spectra. Therefore, compound **11** was tentatively identified as an isomer of compound **1**. Compounds **18**, **27**, **29**, **38**, **40**, **41**, **43** and **47** were identified as 18-hydroxy yohimbine, reserpic acid

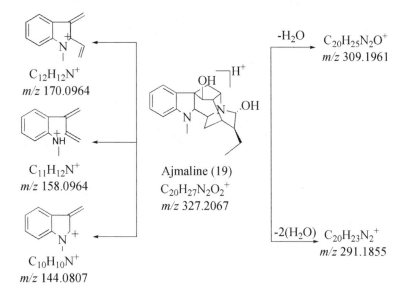

SCHEME 2.1 Proposed fragmentation pathways and diagnostic fragment ions of Ajmaline.

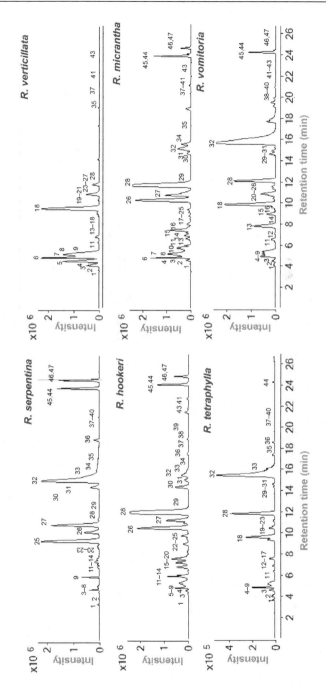

FIGURE 2.3   BPCs of ethanolic extracts of *Rauwolfia* species. (Reproduced from Kumar et al., 2016b with permission from Elsevier.)

TABLE 2.1 Chromatographic and spectrometric characteristics of MIAs in the ethanolic extract of six *Rauvolfia* species (root) by HPLC-ESI-QTOF-MS/MS

S. NO.	RT (MIN)	ERROR (ΔPPM)	OBS. M/Z	MOLECULAR FORMULA	COMPOUNDS	MS/MS FRAGMENT IONS (RELATIVE ABUNDANCE, %)	RS	RV	RH	RM	RT	RVM
									DISTRIBUTION ROOT			
				RESERPINE CLASS								
1	3.40	1.49	341.1867	$C_{20}H_{24}N_2O_3$	Yohimbinic acid	323.1764 (5), 210.1119 (9), 192.1019 (4), 158.0932 (15), 144.0794 (100)	+	+	+	+	+	+
11	6.95	−0.21	341.1865	$C_{20}H_{24}N_2O_3$	Yohimbinic acid (isomer)	323.1762 (3), 210.1153 (11), 192.1018 (5), 158.0953 (15), 144.0797(100),	+	+	+	+	+	+
16	7.81	0.03	401.2071	$C_{22}H_{28}N_2O_5$	Reserpic acid	383.1965 (2), 383.1965 (2), 369.1809 (2), 321.1598 (5), 240.1200 (32), 222.1125 (1), 188.1045 (5), 174.0907 (100)	+	+	+	+	+	+
18	8.20	−0.29	371.1966	$C_{21}H_{26}N_2O_4$	18-Hydroxy-yohmbine	339.1703 (3), 353.1857 (21), 240.1230 (12), 222.1125 (2), 158.0964 (10), 144.0798 (100), 228.1223 (17)	+	+	+	+	+	+
27	10.75	1.44	415.2217	$C_{23}H_{30}N_2O_5$	Reserpic acid methyl ester	383.1916 (4), 397.2122 (2), 254.1369 (51), 236.1281 (2), 222.1116 (7), 188.1059 (7), 174.0899 (100), 160.0743 (4)	−	+	−	+	−	−

(Continued)

TABLE 2.1 (Continued)   Chromatographic and spectrometric characteristics of MIAs in the ethanolic extract of six *Rauwolfia* species (root) by HPLC-ESI-QTOF-MS/MS

S. NO.	RT (MIN)	ERROR (ΔPPM)	OBS. MIZ	MOLECULAR FORMULA	COMPOUNDS	MS/MS FRAGMENT IONS (RELATIVE ABUNDANCE, %)	DISTRIBUTION ROOT					
							RS	RV	RH	RM	RT	RVM
				RESERPINE CLASS								
28	10.91	-0.79	355.2019	$C_{21}H_{26}N_2O_3$	Yohimbine[s]	323.1560 (10), 224.1285(19), 337.1920 (3), 212.1282 (24), 158.0959 (1), 144.0808 (100)	+	+	+	+	+	+
29	11.70	1.44	415.2217	$C_{23}H_{30}N_2O_5$	Seredine	383.1935 (3), 224.1257 (9), 204.1029 (100), 218.1176 (1), 189.0771 (21), 173.0845 (45)	+	+	+	+	+	+
38	20.82	-0.49	595.2651	$C_{32}H_{38}N_2O_9$	Pseudoreserpine	434.1748 (15), 383.1941 (30), 351.1773 (5), 222.1110 (10), 195.0631 (100), 188.10 (3), 174.0880 (43)	+	+	+	+	+	+
39	21.20	0.14	563.239	$C_{31}H_{34}N_2O_8$	Reserpine class (unknown)	351.1711 (60), 333.1535 (42), 319.1463 (55), 222.1244 (32), 195.0672 (21), 170.0939 (34), 144.0800 (100)	+	-	+	+	+	+
40	21.50	0.19	565.2544	$C_{31}H_{36}N_2O_8$	Raunescine	533.2282 (2), 434.1714 (3), 353.1805 (57), 321.1559 (5), 222.1115 (5), 158.063 (2), 195.0622 (100), 144.0777(28),	+	+	-	+	+	+

(Continued)

TABLE 2.1 (Continued) Chromatographic and spectrometric characteristics of MIAs in the ethanolic extract of six *Rauvolfia* species (root) by HPLC-ESI-QTOF-MS/MS

S. NO.	RT (MIN)	ERROR (ΔPPM)	OBS. MIZ	MOLECULAR FORMULA	MS/MS FRAGMENT IONS (RELATIVE ABUNDANCE, %)	COMPOUNDS	RS	RV	RH	RM	RT	RVM
									ROOT			
				RESERPINE CLASS								
41	22.90	0.4	621.2805	$C_{34}H_{40}N_2O_9$	589.2544 (2), 460.1930 (4), 383.1925 (11), 351.1703 (3), 221.0794 (100), 222.1115 (98), 188.1070 (5), 174.0804 (19)	Rescidine	+	+	+	+	–	+
42	23.48	1.45	591.271	$C_{33}H_{38}N_2O_8$	371.1896 (21), 353.1847 (28), 221.0808 (100), 144.0771 (17)	Reserpine class (unknown)	+	–	–	–	–	+
43	23.80	0.03	579.2702	$C_{32}H_{38}N_2O_8$	547.2413 (12), 367.2009 (79), 448.1907 (1), 335.1730 (24), 236.1262 (10), 195.0649 (100), 144.0817 (38)	Deserpidine	+	+	+	+	–	+
44	24.31	–0.25	609.2807	$C_{33}H_{40}N_2O_9$	577.2492 (5), 448.1930 (24), 436.1934 (5), 397.2090 (66), 365.1835 (13), 236.1261 (15), 195.0640 (100), 174.0905 (55)	Reserpine[s]	+	–	+	+	+	+
45	24.54	1.87	607.2634	$C_{33}H_{38}N_2O_9$	367.2017 (17), 335.1765 (11), 236.1272 (7), 221.0820 (100), 206.0577 (10), 190.0621 (14), 144.0815 (35)	Reserpine class (unknown)	+	–	+	+	–	+

(*Continued*)

TABLE 2.1 (Continued)   Chromatographic and spectrometric characteristics of MIAs in the ethanolic extract of six *Rauwolfia* species (root) by HPLC-ESI-QTOF-MS/MS

S. NO.	RT (MIN)	ERROR (ΔPPM)	OBS. M/Z	MOLECULAR FORMULA	MS/MS FRAGMENT IONS (RELATIVE ABUNDANCE, %)	COMPOUNDS	DISTRIBUTION ROOT					
							RS	RV	RH	RM	RT	RVM
				RESERPINE CLASS								
46	25.02	1.78	605.2857	$C_{34}H_{40}N_2O_8$	369.1805 (20), 221.1805 (100), 206.0598 (11), 190.0618 (27), 174.0910 (40)	Reserpine class (unknown)	+	−	+	+	−	+
47	25.22	0.08	635.2963	$C_{35}H_{42}N_2O_9$	603.2779 (1), 474.2066 (5), 397.2085 (22), 236.1281 (2), 221.0791 (100), 188.1070 (2), 174.0881 (11)	Rescinnamine	+	−	+	+	−	+
				Ajmalicine class								
4	5.52	−0.36	353.1861	$C_{21}H_{24}N_2O_3$	321.1598 (1), 222.1124 (1), 210.111 (7), 144.0808 (100), 158.0964 (2)	Ajmalicine[s]	+	+	+	+	+	+
6	6.01	0.05	369.1809	$C_{21}H_{24}N_2O_4$	337.1558 (19), 222.1275 (16), 351.1703 (37), 291.1496 (5), 158.0963 (30), 144.0808 (100)	Ajmalicine derivative	+	+	+	+	+	+
24	10.52	1.07	413.2076	$C_{23}H_{28}N_2O_5$	397.1763 (3), 323.1560 (3), 222.1106 (28), 218.1176 (18), 204.1005 (100)	Reserpiline	+	−	+	+	+	+

(Continued)

TABLE 2.1 (Continued)   Chromatographic and spectrometric characteristics of MIAs in the ethanolic extract of six *Rauvolfia* species (root) by HPLC-ESI-QTOF-MS/MS

S. NO.	RT (MIN)	ERROR (ΔPPM)	OBS. M/Z	MOLECULAR FORMULA	MS/MS FRAGMENT IONS (RELATIVE ABUNDANCE, %)	COMPOUNDS	DISTRIBUTION					
									ROOT			
							RS	RV	RH	RM	RT	RVM
				RESERPINE CLASS								
31	14.81	0.02	413.2072	$C_{23}H_{28}N_2O_5$	397.1766 (10), 323.1560 (2), 222.1125 (29), 204.1005 (100), 218.1176(30)	Reserpiline (isomer)	+	+	+	+	+	+
33	15.50	0.01	383.1965	$C_{22}H_{26}N_2O_4$	351.1677 (2), 222.1110 (10), 159.0675 (3), 188.1053 (2), 174.0895 (100)	10-Demethoxyreserpiline	+	–	+	+	+	+
35	17.50	0.23	411.1914	$C_{23}H_{26}N_2O_5$	379.1652 (1), 222.1110 (16), 204.1007 (100), 218.1181 (12), 173.0818 (11)	Darcyribeirine	+	–	+	+	+	+
Ajmaline class												
2	4.80	0.31	325.1910	$C_{20}H_{24}N_2O_2$	307.1715 (5), 186.0896 (12), 174.0913 (57), 160.0805 (100), 146.0935 (22), 138.0954 (12)	Norseredamine	+	+	+	+	+	+
3	5.01	0.76	311.1752	$C_{19}H_{22}N_2O_2$	293.1752 (2), 172.0763 (9), 160.0768 (21), 146.0960 (100), 136.0791 (10)	Hydroxynortetraphyllicine	+	+	+	+	+	+
8	6.55	1.4	343.2005	$C_{20}H_{27}N_2O_3$	325.1927 (8), 307.1830 (15), 186.0906 (54), 174.0917 (60), 160.0766 (22), 138.0912 (100)	Ajmalinol	+	+	+	+	+	+

(Continued)

TABLE 2.1 (Continued) Chromatographic and spectrometric characteristics of MIAs in the ethanolic extract of six *Rauwolfia* species (root) by HPLC-ESI-QTOF-MS/MS

S. NO.	RT (MIN)	ERROR (ΔPPM)	OBS. M/Z	MOLECULAR FORMULA	MS/MS FRAGMENT IONS (RELATIVE ABUNDANCE, %)	COMPOUNDS	DISTRIBUTION					
									ROOT			
							RS	RV	RH	RM	RT	RVM
				RESERPINE CLASS								
9	6.60	0.09	313.191	$C_{19}H_{24}N_2O_2$	295.1806 (5), 196.1123 (21), 156.0813 (16), 144.0821 (55), 130.0664 (100)	Norajmaline	+	+	+	+	+	+
13	7.45	1.52	503.2386	$C_{26}H_{34}N_2O_3$	341.1898 (10), 323.1752 (62), 309.1595 (27), 291.1487 (100)	Hydroxyseredamine-O-hexoside	+	−	+	+	+	+
14	7.52	1.26	295.1801	$C_{19}H_{22}N_2O$	277.1666 (4), 156.0799 (7), 144.0811 (27), 138.0474 (47), 130.0656 (100), 120.0791 (24)	Nortetraphyllicine	+	+	+	+	+	+
19	8.91	0.13	327.2067	$C_{20}H_{26}N_2O_2$	309.1991 (3), 291.1839 (3), 170.0974 (45), 158.0964 (100), 144.0812 (82)	Ajmaline[s]	+	+	+	+	+	+
21	9.53	0.6	309.196	$C_{20}H_{24}N_2O$	291.1780 (4), 263.1536 (31), 170.0950 (19), 158.0965 (77), 144.0817 (100), 120.0822	Tetraphyllicine	+	−	+	+	+	−
22	9.84	1.4	353.186	$C_{21}H_{24}N_2O_3$	335.1758 (3), 275.1501 (2), 156.0808 (4), 144.0890 (100), 130.0693 (7)	Quebrachidine	+	+	+	+	+	+
23	10.11	0.11	355.2017	$C_{21}H_{26}N_2O_3$	309.2031 (5), 277.1667 (13), 260.1360 (12), 170.0955 (78), 158.0956 (98), 144.0805 (78)	Hydroxymethylsere-damine	+	+	+	+	+	+

(Continued)

TABLE 2.1 (Continued)   Chromatographic and spectrometric characteristics of MIAs in the ethanolic extract of six *Rauwolfia* species (root) by HPLC-ESI-QTOF-MS/MS

S. NO.	RT (MIN)	ERROR (ΔPPM)	OBS. M/Z	MOLECULAR FORMULA	MS/MS FRAGMENT IONS (RELATIVE ABUNDANCE, %)	COMPOUNDS	DISTRIBUTION					
							ROOT					
							RS	RV	RH	RM	RT	RVM
RESERPINE CLASS												
36	19.31	0.28	351.2066	$C_{22}H_{26}N_2O_2$	291.1827 (17), 182.962 (78), 170.0956 (45), 158.0962 (45), 144.0803 (100), 120.0802 (10)	17-O-Acetyltetra-phyllicine	+	–	+	–	+	–
37	20.76	0.45	489.2595	$C_{26}H_{36}N_2O_7$	327.2073 (100), 309.1663 (80), 291.1554 (47)	Ajmaline-O-hexoside	–	+	+	+	–	–
Quaternary alkaloids												
15	7.70	1.25	513.2232	$C_{27}H_{32}N_2O_8$	351.1674 (4), 333.1615 (14), 319.1440 (3), 274.1199 (16), 246.1299 (53), 230.0986 (87), 220.1137 (67), 204.0823 (100), 195.0926 (65)	Tetradehydroyohimbine-O-hexoside	+	+	+	+	+	+
20	9.17	0.25	511.2373	$C_{37}H_{30}N_2O_8$	349.1548 (100), 317.1326 (24), 289.1313 (10)	Serpentine-O-hexoside	+	+	+	+	+	+
25	10.60	0.85	335.1389	$C_{20}H_{18}N_2O_3$	317.1296 (100), 289.1492 (7), 263.0808 (71), 235.0863 (34), 307.0858 (5)	Serpentine derivative	+	+	–	+	+	+
26	10.70	–1.54	351.1707	$C_{21}H_{22}N_2O_3$	351.1674 (100), 333.1570 (4), 219.1440 (5), 271.1470 (20), 248.1024 (4), 195.0890 (6), 168.0788 (13)	Tetradehydroyohimbine	+	+	+	+	+	+

(Continued)

TABLE 2.1 (Continued) Chromatographic and spectrometric characteristics of MIAs in the ethanolic extract of six *Rauwolfia* species (root) by HPLC–ESI-QTOF-MS/MS

S. NO.	RT (MIN)	ERROR (ΔPPM)	OBS. M/Z	MOLECULAR FORMULA	MS/MS FRAGMENT IONS (RELATIVE ABUNDANCE, %)	COMPOUNDS	RS	RV	RH	RM	RT	RVM
							\[DISTRIBUTION — ROOT\]					
				RESERPINE CLASS								
32	15.40	0.47	349.1547	$C_{21}H_{20}N_2O_3$	317.1290 (66), 289.1314 (7), 263.0796 (100), 235.0826 (34), 307.0858 (6)	Serpentine[s]	+	−	+	+	+	+
34	16.62	0.45	685.3375	$C_{42}H_{45}N_4O_5$	653.3131 (13), 435.1947 (23), 403.1516 (3), 349.1235 (2), 375.1701 (23), 251.1558 (7)	Serpentinine	+	−	+	+	−	−
				Other compounds								
5	5.93	1.47	343.2015	$C_{20}H_{26}N_2O_3$	325.1972 (47), 307.1794 (16), 182.0968 (57), 170.0945 (36), 158.0962 (81), 144.0795 (100), 132.0757 (11)	Rauvotetraphylline A	+	+	+	+	+	+
7	6.33	1.25	313.1917	$C_{19}H_{24}N_2O_2$	295.1845 (7), 144.0819 (70), 138.0933 (100), 122.0869 (61)	Dihydroperaksine	+	+	−	+	+	+
10	6.90	0.89	311.1755	$C_{19}H_{22}N_2O_2$	293.1641 (8), 156.0808 (17), 144.0804 (52), 138.0919 (10), 130.0650 (100), 120.0805 (28)	Norajmalidine	+	−	−	+	+	−

(Continued)

TABLE 2.1 (Continued)   Chromatographic and spectrometric characteristics of MIAs in the ethanolic extract of six *Rauvolfia* species (root) by HPLC-ESI-QTOF-MS/MS

S. NO.	RT (MIN)	ERROR (ΔPPM)	OBS. M/Z	MOLECULAR FORMULA	MS/MS FRAGMENT IONS (RELATIVE ABUNDANCE, %)	COMPOUNDS	DISTRIBUTION ROOT					
							RS	RV	RH	RM	RT	RVM
				RESERPINE CLASS								
12	7.12	0.12	513.2231	$C_{27}H_{32}N_2O_8$	351.1994 (100), 291.1480 (16)	Raucaffricine	+	+	+	-	+	+
17	8.02	0.15	309.1598	$C_{19}H_{20}N_2O_2$	291.1492, 172.0782 (24), 160.0757 (51), 146.0612 (100) , 172.0782 (24), 136.0780 (38), 120.0803(15)	Normitoridine	+	+	+	+	+	+
30	12.04	0.74	429.2018	$C_{23}H_{28}N_2O_6$	397.1781 (87), 369.1759 (10), 353.1501 (3) 339.1340 (14), 220.0972 (100), 210.1124 (35), 205.0732 (9), 189.0772 (21), 178.0845 (13), 150.0913 (21)	Carapanaubine	+	+	+	+	-	+

Rs: R. serpentine; Rv: R. verticillata; Rh: R. hookeri; Rm: R. micrantha; Rt: R. tetraphylla; and *Rvm: R. vomitoria.*
Source: Reproduced from Kumar et al., 2016b with permission from Elsevier.

methyl ester, seredine, pseudoreserpine, raunescine, rescidine, deserpidine and rescinnamine, respectively.

The molecular mass of compound **18** (*m/z* 371.1966) was 16 Da more than compound **28**, showing the presence of an extra OH group. This was also corroborated by the C-ring cleavage peak at *m/z* 240.1230, which is again 16 Da more than that in compound **28**. Therefore, compound **18** was identified as 18-hydroxy yohimbine. The molecular mass of compound **27** (*m/z* 415.2217) was 14 Da more than that of compound **16**. The same difference was also observed in the C-ring cleavage peak at *m/z* 254.1369, indicating the presence of a methyl ester group. The indole fragment peak was unchanged at *m/z* 174.0899. Compound **27** was therefore identified as reserpic acid methyl ester. There is a 60-Da difference between the molecular masses of compound **29** (*m/z* 415.2217) and compound **28**, indicating the presence of two methoxy groups in compound **27**. The C-ring cleavage peaks at *m/z* 224.1257 (same as in compound **28**) and 204.1029 (60 Da more than that in compound **28**) suggest that the two methoxy groups are in the indole part of the molecule. Hence, compound **29** was identified as seredine. The molecular mass of compound **38** (*m/z* 595.2651) was 14 Da less than that of reserpine. The C-ring cleavage showed fragment ions at *m/z* 434.1748 and *m/z* 174.0880 (same as in compound **44**), indicating that the difference in substitution is in the terpene moiety in compound **38**. The presence of 3,4,5-trimethoxy benzoyl group is confirmed by the loss of 3,4,5-trimethoxy benzoic acid from *m/z* 595.2651 and 434.1714, resulting in the ions at *m/z* 383.1941 and 222.1110, respectively, and the most abundant ion at *m/z* 195.0631 in compound **38**. Compound **38** was, therefore, identified as pseudoreserpine. The difference of 30 Da in the molecular masses of **38** and **40** (*m/z* 565.2544) shows that compound **40** has one methoxy group less. The C-ring cleavage gave fragment ions at *m/z* 434.1714 and *m/z* 144.0777, suggesting that the difference is in the indole part of the molecule in compound **40**. The presence of 3,4,5-trimethoxybenzoyl ion (*m/z* 195.0622, 100%) and the losses of 3,4,5-trimethoxy benzoic acid (*m/z* 353.1805 and 222.1115) confirm the 3,4,5-trimethoxy benzoyl substitution. Hence, compound **40** was identified as raunescine. The molecular mass of **41** (*m/z* 621.2805) indicates a difference of 12 Da with that of compound **44**. The same difference is also observed in the C-ring cleavage product at *m/z* 460.1930, suggesting a difference in substitution in the terpene moiety. This is also indirectly confirmed by the no change observed in the indole product ion of the C-ring cleavage (*m/z* 174.0804). The observed losses of 3,4,5-trimethoxycinnamic acid from [M+H]⁺ and C-ring cleavage product at *m/z* 460.1930 resulting in the ions at *m/z* 383.1925 and 222.1115, and the presence of the most abundant peak corresponding to 3,4,5-trimethoxycinnamoyl ion at *m/z* 221.0794 lead to the identification of compound **41** as rescidine. The molecular mass of compound **43** (*m/z* 579.2702) is 30 Da less than that of compound **44**, i.e., one methoxy group less. Comparing the spectra of compounds **43** and **44**, it is clear that the difference is in the indole product ion of the C-ring cleavage at m/z 144.0817 against m/z 174.0905 in compound **44**, thereby suggesting that **43** is deserpidine. Similarly, for **47**, the molecular mass difference with reserpine is 26 Da, the difference between 3,4,5-trimethoxybenzoic acid and 3,4,5-trimethoxycinnamic acid. This difference is seen in all the product ions containing this group. All the other product ions remain the same. Hence, the compound **47** is identified as rescinnamine.

2.3.3.2 Ajmalicine Class of Compounds

The fragmentation of ajmalicine is similar to those of yohimbine and reserpine (Figure 2.2). Six compounds **4, 6, 24, 31, 33** and **35** were identified and characterized as ajmalicine, ajmalicine derivative, reserpiline, reserpiline (isomer), 10-demethoxyreserpiline and darcyribeirine, respectively. Elimination of CH_3OH from compounds **6, 24, 33** and **35** gave fragment ions at m/z 337.1558, 323.1560, 351.1677 and 379.1652, respectively. Fragment ions at m/z 158.0963 (compound **6**), 218.1174 (compounds **24** and **35**) and 188.1053 (compound **33**) were observed due to the loss of terpene moiety via RDA cleavage followed by bond cleavage between C14 and C15. Similarly, the fragment ion at m/z 144.0808 (compound **6**) suggests no methoxy group, that at m/z 174.0895 (compound **33**) suggests one methoxy group, and that at m/z 204.1005 (compounds **24** and **35**) suggest two methoxy groups on the indole moiety in these compounds. Compounds **24** and **31** may be an isomeric pair, showing similar fragments, but with different relative abundance eluting at 10.52 min and 14.81 min, respectively.

2.3.3.3 Ajmaline Class of Compounds

Successive losses of H_2O molecules and ring cleavages resulting in N-methyl indole fragments are the major fragmentations of ajmaline (Scheme 2.1). Ajmaline (**19**) was confirmed by comparison with the standard. Eleven more ajmaline class of compounds **2, 3, 8, 9, 13, 14, 21, 22, 23, 36** and **37** were tentatively identified and characterized as seredamine, hydroxynortetraphyllicine, ajmalinol, norajmaline, hydoxyseredamine-O-hexoside, nortetraphyllicine, tetraphyllicine, quebrachidine, hydroxymethylseredamine, 17-O-acetyltetraphyllicine and ajmaline-O-hexoside, respectively. Compounds **2, 3, 8, 9, 14, 21, 22, 23** and **36** showed fragment ions at m/z 307.1715, 293.1752, 325.1927, 295.1806, 277.1666, 291.1780, 335.1758, 309.2031 and 291.1827 due to the loss of H_2O except compound **36**, which showed the loss of acetic acid. Fragment ions at m/z 186.0896 (compound **2**), 172.0763 (compounds **3** and **8**), 156.0813 (compounds **9, 14** and **22**) and 170.0974 (compounds **21, 23** and **36**) were observed due to the loss of terpene moiety via ring cleavage reactions. Similarly, fragment ions at m/z 174.0913 (compound **2**), 160.0768 (compounds **3** and **8**), 144.0821 (compounds **9**, **14** and **22**) and 158.09 (compounds **21**, **23** and **36**) are formed due to the loss of terpene moiety via ring cleavage reactions. Other ring cleavage fragment ions arising from the loss of terpene moiety were at m/z 160.0805 (compounds **2** and **8**), 146.0960 (compound **3**), 130.0656 (compounds **9**, **14** and **22**) and 144.0817 (compounds **21**, **23** and **36**). Compounds **13** and **37** were identified as hydoxyseredamine-O-hexoside and ajmaline-O-hexoside, which showed a characteristic loss of hexoside (162 Da) and gave fragment ions at m/z 341.1898 and 327.2073, respectively.

2.3.3.4 Quaternary Indole Alkaloids

The fragmentation of serpentine is probably initiated by RDA cleavage of ring E followed by elimination of CH_3OH and $C_5H_{10}O$, resulting in the ions at m/z 263.0796 and 317.1290, respectively. Further fragmentations involve the loss of CO. Compound 32 eluting at 15.40 min was confirmed as serpentine by comparison with the standard. Five more quaternary indole alkaloid compounds **15**, **20**, **25**, **26** and **34** were identified and characterized as tetradehydroyohimbine-O-hexoside, serpentine-O-hexoside, serpentine derivative, tetradehydroyohimbine and serpentinine, respectively. Compound **15** (m/z 513.2232) showed fragment ion at m/z 351.1674 due to the loss of anhydrohexose ($C_6H_{10}O_5$, 162 Da). This is the behavior expected of a hexoside molecule, and hence, it was identified as tetradehydroyohimbine-O-hexoside. The fragment ion at m/z 351.1674 gave product ions at m/z 333.1615 and 319.1440 due to further losses of H_2O and CH_3OH, respectively. Compound **20** showed fragment ion at m/z 349.1548 due to the characteristic loss of anhydrohexose. Further loss of CH_3OH resulted in the ion at m/z 317.1326. Compound **25** was identified as a serpentine derivative (m/z 335.1389), which showed a molecular mass 14 Da lower than that of serpentine, but followed a similar fragmentation pattern. It showed fragment ions at m/z 317.1296 and 289.1492 due to the losses of H_2O and CO, respectively. Compound **34** (m/z 685.3375) showed fragment ions at m/z 653.3131, 435.1956 and 251.1546 due to the losses of CH_3OH, $C_{17}H_{18}N_2$ and $C_{25}H_{26}N_2O_5$ respectively. Fragment ion m/z 435.1956 showed product ions at m/z 403.1516, 375.1732 and 349.1235 due to the losses of CH_3OH, $C_2H_4O_2$ and $C_4H_6O_2$, respectively.

2.3.3.5 Other Indole Alkaloids

Six indole alkaloids **5**, **7**, **10**, **12**, **17** and **30** were tentatively identified and characterized as rauvotetraphylline A, dihydroperaksine, norajmalidine, raucaffricine, normitoridine and carapanaubine, respectively. Compounds **5**, **7**, **10** and **17** showed fragment ions at m/z 325.1972, 295.1845, 293.1641 and 291.1492, respectively, due to the loss of H_2O from the respective precursor ions at *m/z* 343.2015, 313.1917, 311.1755 and 309.1598. The product ions at *m/z* 158.0962 [143+CH_3], 144.0795, 156.0808 and 172.0782 were observed in compounds **5, 7, 10** and **17**. Similarly, fragment ions at *m/z* 130.0650 and 146.0612 [129+OH] were observed as base peaks in compounds **10** and **17**, respectively. Compound **12** showed fragment ions at *m/z* 351.1994 due to the loss of $C_6H_{10}O_5$. Compound **30** showed fragment ions at *m/z* 397.1781, 369.1759 and 353.1501 due to successive losses of CH_3OH, CO and CH_4, respectively. Ring cleavages resulted in the fragment ions at *m/z* 210.1124 and 220.0972, which on subsequent fragmentation resulted in the ions at *m/z* 178.0845 and 150.0913 due to the losses of CH_3OH and CO from *m/z* 210.1124 and *m/z* 205.0732 and 189.0772 due to the losses of CH_3 and CH_4 from *m/z* 220.0972.

2.3.4 Identification of Markers Using PCA

PCA converts a large number of data sets to a much smaller number of variables. It produces overall discrimination between the closely related samples for quality control and authentication (Ian, 2002; Bajpai et al., 2015; D'Urso et al., 2015). LC-MS data, in combination with a data reduction technique such as PCA, serves as an efficient and powerful tool to identify the chemical markers (Kumar et al., 2015; Bajpai et al., 2015; D'Urso et al., 2015). The LC-MS chemical fingerprints of *R. serpentina*, *R. verticillata*, *R. hookeri*, *R. micrantha*, *R. tetraphylla* and *R. vomitoria* roots were analyzed by PCA to identify the chemical markers for discrimination among these species. A total of 66 peaks from mass range *m/z* 179.0708 to 635.2984, peak area ≥1,000, were taken from the HPLC-ESI-QTOF-MS fingerprints ($n = 3$) of *Rauvolfia* roots for running the PCA. The PC1 and PC2 together were able to explain 57.96% of variance information. To obtain the best expression, peaks with lowest contribution were dropped and only 12 peaks at *m/z* 323.2134 (unknown), 325.1910 (seredamine), 343.2005 (ajmalinol), 327.2067 (ajmaline), 327.2104 (isomer of ajmaline), 355.2019 (yohimbine), 355.2029 (isomer of yohimbine), 349.1547 (serpentine), 383.1965 (10-demethoxyreserpiline), 351.1707 (tetradehydroyohimbine), 413.2067 (reserpiline) and 621.2805 (rescidine) were identified as marker peaks, which were responsible for discrimination of six *Rauvolfia* species. Using these chemical markers, PCs could explain 80.03% of the variance information as shown in the score and loading plots (Figures 2.4a and b). Peak at *m/z* 343.2005 (46.55%) showed a higher contribution followed by *m/z* 327.2104 (33.47%) and 327.2067 (12.28%), respectively. Similar report with contribution of peak area has already been available in chromatographic fingerprinting and quantitative analysis of Xinkeshu tablet (Wang et al., 2012). The PCA plots afforded useful qualitative information on the roots of six *Rauvolfia*

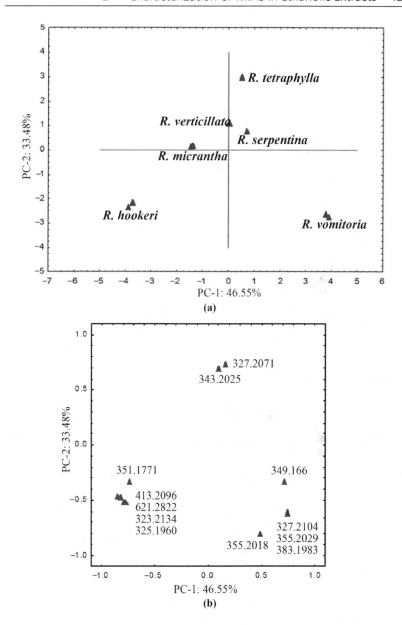

FIGURE 2.4 PCA scores plot (a) from *Rauwolfia* species showing discrimination among the *R. serpentina*, *R. verticillata*, *R. hookeri*, *R. micrantha*, *R. tetraphylla* and *R. vomitoria*. Loadings plots (b) of the normalized LC-MS data obtained from three repeats. (Reproduced from Kumar et al., 2016b with permission from Elsevier.)

species and revealed similarities and dissimilarities among them as shown in Figure 2.4a. It showed that *R. serpentina*, *R. micrantha* and *R. verticillata* were close, whereas *R. hookeri*, *R. tetraphylla* and *R. vomitoria* were much apart from each other. *R. serpentina* is commonly used in herbal formulations and endangered due to overexploitation (Sharma and Chandel, 1992). This study showed that *R. micrantha* and *R. verticillata* might be used as substitutes for *R. serpentina* in herbal formulations. Hence, this analysis will also help to use an alternative plant based on the phytochemical investigation, which may conserve *R. serpentina*.

2.4 CONCLUSIONS

A simple, sensitive, reproducible and accurate HPLC-ESI-QTOF-MS/MS method was developed in positive-ion mode for the dereplication of MIAs. Fragmentation pathways were established with the help of MS/MS spectra of standard compounds. Reserpine, ajmalicine, ajmaline and yohimbine classes of MIAs showed two types of fragment ions, namely those due to loss of substituents that were attached with the terpene moiety and C-ring cleavages. Based on these diagnostic fragmentation pathways, the MIAs in the extracts were characterized and identified. In total, 47 compounds were tentatively identified in ethanolic extract of the roots of *R. serpentina*, *R. verticillata*, *R. hookeri*, *R. micrantha*, *R. tetraphylla* and *R. vomitoria*. Hydoxyseredamine-*O*-hexoside (13), ajmaline-*O*-hexoside (**37**), tetradehydroyohimbine-*O*-hexoside (**15**), serpentine-*O*-hexoside (**25**) and four reserpine class compounds 39, 42, 45 and 46 were identified as unknown/new for the first time. The isomeric compounds reserpic acid methyl ester (**27**) and seredine (**29**) were successfully distinguished by MS/MS analysis. The marker peaks were successfully identified by PCA, which can discriminate *R. serpentina*, *R. verticillata*, *R. hookeri*, *R. micrantha*, *R. tetraphylla* and *R. vomitoria* for quality control and authentication. This report also emphasizes the importance of accurate mass measurements as it speeds up the dereplication process.

Simultaneous Determination of Bioactive Monoterpene Indole Alkaloids

3

3.1 INTRODUCTION

As a part of the quality control program on herbal raw materials and for-mulations, a fast and reliable ultra-performance liquid chromatography (UPLC)-triple quadrupole linear ion trap MS method was developed for the simultaneous quantification of the major monoterpene indole alkaloids. Among the *Rauvolfia* species, *R. serpentina* is the most common one used in India. Because of the indiscriminate use of this plant, it is now an endan-gered species, and the other species found in India, namely, *Rauvolfia hookeri*, *R. micrantha*, *R. tetraphylla*, *R. verticillata* and *R. vomitoria*, are also used as substitutes. To balance the efficacies of the herbal products, it is essential to have information about the contents of the various bioactive components in these species as well. It is well known that indole alkaloids are the major bioactive components in *Rauvolfia*. It was, therefore, decided to quantitatively estimate the major alkaloids ajmaline, yohimbine, ajmalicine, serpentine and reserpine in the ethanol extracts obtained from the leaves and roots of *R. serpentina*, *R. verticillata*, *R. hookeri*, *R. micrantha*, *R. tetraphylla* and *R. vomitoria* using an ultra-performance liquid chromatography-electrospray ionization-tandem mass spectrometry (UPLC-ESI-MS/MS) method in multiple reaction monitoring (MRM) acquisition mode.

3.2 SAMPLE PREPARATION

Liquid chromatography-mass spectrometry (LC-MS)-grade methanol, acetonitrile and formic acid were purchased from Sigma-Aldrich (St Louis, MO, USA). AR-grade ethanol purchased from Merck Millipore (Darmstadt, Germany) was used in preparing the extracts. Ultra-pure water obtained from Direct-Q system (Millipore, Billerica, MA, USA) was used throughout the analysis. The standard compounds ajmaline (10 mg, purity ≥98%), yohimbine (10 mg, purity ≥98%), ajmalicine (10 mg, purity ≥98%), serpentine (5 mg, purity ≥98%) and reserpine (25 mg, purity ≥99%) were purchased from Sigma-Aldrich (St Louis, MO, USA) and from ChemFaces (Wuhan, Hubei, China). The roots and leaves of *R. serpentina*, *R. verticillata*, *R. hookeri*, *R. micrantha*, *R. tetraphylla* and *R. vomitoria* were collected from the plants grown under similar conditions in Jawaharlal Nehru Tropical Botanic Garden and Research Institute (JNTBGRI) campus, Kerala, India and the extracts were prepared as described in Section 2.2.2. Four herbal formulations (HF1, HF2, HF3 and HF4) of *Rauvolfia serpentina* produced by different Indian herbal drug companies were purchased from local drug stores in Lucknow, India. Stock solutions (1,000 µg/mL) of standards ajmaline (1), yohimbine (2), ajmalicine (3), serpentine (4) and reserpine (5) in methanol were diluted in 13 concentrations (0.5, 1.0, 2.0, 2.5, 5.0, 10, 20, 25, 50, 100, 200, 250 and 500 µg/mL). The calibration curves were constructed by plotting the values of peak areas versus the values of concentrations of each analyte. All stock solutions were stored at −20°C until analysis. An aliquot of the *Rauvolfia* ethanolic extract (approximately 1 mg) was dissolved in methanol (100 mL), sonicated for 30 min and filtered through a 0.22-µm syringe filter (Millex-GV, PVDF, Merck Millipore, Darmstadt, Germany). The filtrates were further diluted with methanol and vortexed for 5 min to obtain concentrations in the range of 100 to 1,000 ppb. After completely removing the coating of herbal formulation, a subsample (approximately 0.5 g) of each sample was placed in methanol (50 mL) at 26°C–28°C and sonicated for 30 min. After centrifugation at 15,000 rpm for 10 min, the supernatant was filtered through a 0.22-µm syringe filter (Millex-GV, PVDF, Merck Millipore, Darmstadt, Germany). The filtrates were diluted with methanol to prepare working concentrations for the analysis.

3.3 LC-QTRAP-MS ANALYSIS

3.3.1 Chromatographic Conditions

Chromatographic separation of MIAs was achieved with an ACQUITY UPLC BEH™ C18 column (1.7 µm, 2.1 mm × 50 mm) operated at 25°C. The mobile phase, which consisted of a 0.1% formic acid in water (A) and acetonitrile (B), was delivered at a flow rate of 0.3 mL/min under a gradient program: 0–1.0 min, 10–10% (B); 1.0–2.0 min, 10–20% (B); 2.0–3.0 min, 20–22% (B); 3.0–5.0 min, 22–25% (B); 5.0–5.20 min, 25–26% (B); 5.20–5.50 min, 26–98%; 5.5–6.0 min, 98–98% (B); 6.0–6.5 min, 98–10% (B); and 6.5–7.0 min, 10–10% (B). The sample injection volume was 2.0 µL.

3.3.2 Mass Spectrometer Conditions

The triple quadrupole linear ion trap mass spectrometer (API 4000 QTRAP™ MS/MS system from AB Sciex, Concord, ON, Canada) equipped with an electrospray (Turbo V™) ion source and coupled with the UPLC system was operated in positive ionization mode. A Turbo ionspray® probe was vertically positioned 11 mm from the orifice and charged with 5,500 V. Source-dependent parameters such as temperature and gas pressure for gas 1, gas 2 and curtain gas were set at 550°C, 50, 50 and 20 psi, respectively. The collision-activated dissociation gas was set as medium, and the interface heater was on. Nitrogen was used as the gas for all the processes. Quadrupole 1 and quadrupole 2 were maintained at unit resolution. Optimization of the mass spectrometric conditions was carried out by infusing (10 µL/min) solutions of the analytes dissolved in methanol (50 ng/mL) using a Harvard "22" syringe pump (Harvard Apparatus, South Natick, MA, USA). For the MRM quantitation, the highest abundance of precursor-to-product ions for each compound was selected (Table 3.1). The dwell time for both precursor and product ions were set at 200 ms. Full-scan ESI-MS spectra were recorded from m/z 100–1,000. AB Sciex Analyst software version 1.5.1 was used to control the LC-MS/MS system and for data acquisition and processing. All the statistical calculations

TABLE 3.1   MRM parameters for quantitative analysis

ANALYTES	RT (MIN)	PRECURSOR ION	Q3 MASS (PRODUCT ION)		DP	EP	CE	CXP
			QUANTIFIER	QUALIFIER				
		M/Z			(EV)			
Ajmaline	2.41	327 [M+H]+	144	158.3	141	9.0	50	4.2
Yohimbine	2.62	355 [M+H]+	144	212.1	144	9.0	34	13.0
Ajmalicine	3.84	353 [M+H]+	144	178.2	120	6.0	33	22.0
Serpentine	4.14	349 [M]+	317	262.9	128	5.7	34	15.0
Reserpine	5.78	609 [M+H]+	195	173.9	140	7.8	40	32.0

Source: Reproduced from Kumar et al., 2016c with permission from John Wiley & Sons.

related to quantitative analysis were performed using Graph Pad Prism software version 5. The principal component analysis (PCA) was performed using STATISTICA software, Windows version 7.0 (Stat Soft, Inc., Tulsa, OK, USA). The contents of five MIAs in the root and leaf parts of six *Rauvolfia* species were determined simultaneously, and three replicates of each were used for PCA.

3.3.3 Optimization of Analysis

The MS/MS spectra showed fragment ions at m/z 144 (ajmaline, yohimbine and ajmalicine), 317 (serpentine) and 195 (reserpine) as base peaks that were hence selected as quantifiers (Table 3.1). Accompanying ions at m/z 158, 212, 178, 263 and 174 were selected as qualifiers. Compound-dependent MRM parameters such as declustering potential (DP), entrance potential (EP), collision energy (CE) and cell exit potential (CXP) were adjusted to achieve the most specific and stable MRM transitions (precursor-to-product ions) in positive ion mode (Table 3.1). MS/MS spectra and extracted ion chromatograms of compounds are given in Figure 3.1.

3.3.4 Analytical Method Validation

The developed UPLC-ESI-MS/MS method for quantitative analysis was validated according to the 2005 International Conference on Harmonisation (ICH) Guidelines by determining specificity, linearity, lower limit of detection (LOD), lower limit of quantitation (LOQ), precision, solution stability

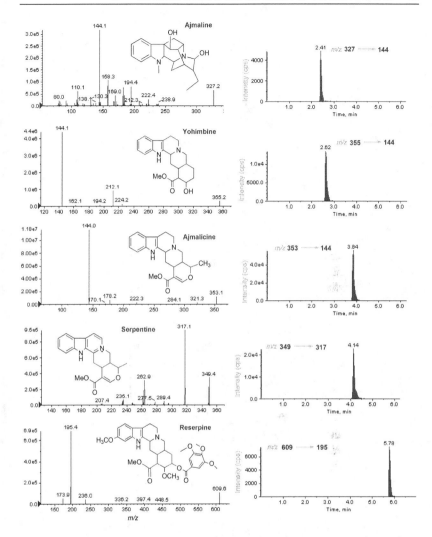

FIGURE 3.1 MS/MS spectra and extracted ion chromatograms of compounds ajmaline, yohimbine, ajmalicine, serpentine and reserpine. (Reproduced from Kumar et al., 2016c with permission from John Wiley & Sons.)

and recovery (Table 3.2). The UPLC-ESI-MS/MS method was employed to analyze two MRM signals (quantifier and qualifier) for each sample. All the peaks of the targeted analytes in *Rauvolfia* species (root and leaf) were unambiguously identified by comparison of retention time, quantifier and qualifier transitions in the MRM chromatograms of standards.

TABLE 3.2 Regression equations, correlation coefficients, linearity ranges, LOD, LOQ, intraday, interday precisions, stability and recovery for ajmaline, yohimbine, ajmalicine, serpentine and reserpine

ANALYTES	REGRESSION EQUATION	R^2	LINEAR RANGE (NG/ML)	LOD	LOQ	PRECISION RSD (%) INTRADAY (N = 6)	INTERDAY (N = 3)	STABILITY RSD % (N = 5)	RECOVERY RSD (%)
Ajmaline	$y = 652.3x - 14.58$	0.999	0.5–250	0.07	0.22	1.60	1.38	1.53	1.82
Yohimbine	$y = 2691x - 125$	1.000	0.5–100	0.15	0.44	0.53	0.83	0.71	2.23
Ajmalicine	$y = 11623x - 218.0$	0.999	0.5–25	0.06	0.19	1.61	2.74	0.12	1.75
Serpentine	$y = 4592x + 273.4$	0.999	0.5–100	0.13	0.41	0.92	1.42	1.21	0.12
Reserpine	$y = 977.2x - 17.48$	0.999	0.5–250	0.06	0.18	2.24	0.76	0.09	0.45

Source: Reproduced from Kumar et al., 2016c with permission from John Wiley & Sons.

3.3.4.1 Linearity, Limit of Detection (LOD) and Limit of Quantitation (LOQ)

Linearity, LOD and LOQ of the five targeted MIAs were determined by serial dilution of sample solutions using the described UPLC-ESI-MS/MS method. Good linear relationship between the peak areas and concentrations was obtained for each of the five analytes over the tested concentration range with a correlation coefficient (r^2) of 0.9996 to 1.000. The LOD (0.06–0.15 ng/mL) and LOQ (0.18–0.44 ng/mL) under the present chromatographic conditions were determined on the basis of response and slope of each regression equation at S/N ratio 3:1 and 10:1, respectively.

3.3.4.2 Precision, Stability and Accuracy

Intraday and interday variations of the method were determined by analyzing known concentrations of the five MIAs in six replicates during a single day and by triplicating the experiments on five successive days, respectively. The percentage relative standard deviation (%RSD) values for precision were in the range of 0.53–2.24% and 0.76–2.74% for intraday and interday assays, respectively. To evaluate the reproducibility of the developed method, six replicates of the same sample with the same concentration (500 ppb) of *R. serpentina* root and leaf were analyzed. The %RSD (≤3.1%) of the five compounds showed high reproducibility of the method. Stability of standard solutions stored at 26°C–28°C was investigated by replicate experiments at 0, 2, 4, 8, 12 and 24 h. The %RSD values of stability were ≤1.53%. To evaluate the accuracy of this method, a recovery test was performed using the standard addition method. The mixed standard solutions with three different spike levels (low, middle and high) were added into a sample and analyzed in triplicate. The recovery was calculated by the formula: recovery = $(a - b)/c \times 100\%$, where "a" is the detected amount, "b" is the original amount and "c" is the spiked amount. The results showed that the developed analytical method was reliable and reproducible with a good recovery in the range 92.2–102.7% (%RSD ≤ 2.23%) for all analytes.

3.4 QUANTITATIVE ANALYSIS OF BIOACTIVE ALKALOIDS IN *RAUVOLFIA* SPECIES

The developed UPLC-ESI-MS/MS method was applied to the ethanolic extracts of roots and leaves of *R. serpentina*, *R. verticillata*, *R. hookeri*,

R. micrantha, *R. tetraphylla* and *R. vomitoria* and herbal formulations of *R. serpentina* for quantitative analysis of the five targeted MIAs (Figure 3.2, Table 3.3).

Almost similar content of ajmaline was detected in the roots of *Rauvolfia serpentina* (52.27 mg/g) followed by *R. vomitoria* (48.43 mg/g) and *R. verticillata* (47.21 mg/g). The root of *R. tetraphylla* showed highest content of yohimbine (100.21 mg/g) and ajmalicine (120.51 mg/g). Similarly, the maximum content of serpentine was detected in the roots of *R. vomitoria* (87.77 mg/g) followed by *R. serpentina* (76.38 mg/g). Approximately the same quantity of reserpine was found in the roots of *R. serpentina* (35.18 mg/g) and *R. micrantha* (32.38 mg/g) followed by *R. hookeri* (23.44 mg/g). It was only 21.14 mg/g

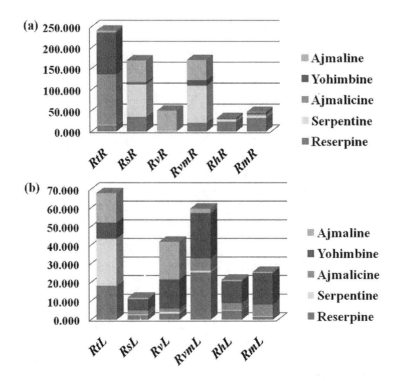

FIGURE 3.2 Graphical representation of contents (mg/g) of ajmaline, yohimbine, ajmalicine, serpentine and reserpine in root (a) and leaf (b) of *R. tetraphylla* Root (*RtR*), *R. verticillata* Root R (*RvR*), *R. vomitoria* Root (*RvmR*), *R. hookeri* Root (*RhR*), *R. micrantha* Root (*RmR*), *R. serpentina* Root (*RsR*), *R. tetraphylla* Leaf (*RtL*), *R. verticillata* Leaf R (*RvL*), *R. vomitoria* Leaf (*RvmL*), *R. hookeri* Leaf (*RhL*), *R. micrantha* Leaf (*RmL*), and *R. serpentina* Leaf (*RsL*). (Reproduced from Kumar et al., 2016c with permission from John Wiley & Sons.)

3 • Simultaneous Determination of Bioactive MIAs   53

TABLE 3.3 The content (mg/g) of compounds ajmaline, yohimbine, ajmalicine, serpentine and reserpine in ethanolic extracts of roots and leaves of *Rauwolfia* species and herbal formulations

METABOLITES	SPECIES NAME	AJMALINE	YOHIMBINE	AJMALICINE	SERPENTINE	RESERPINE
PLANT PARTS				QUANTITY (MG/G)		
Root	R. serpentina	52.27	3.11	3.14	76.38	35.18
	R. verticillata	47.21	0.23	0.12	1.82	0.96
	R. hookeri	0.50	2.72	0.30	5.89	23.44
	R. micrantha	1.72	2.73	4.28	6.01	32.38
	R. tetraphylla	5.07	100.21	120.51	1.51	14.45
	R. vomitoria	48.43	11.63	2.12	87.77	21.14
Leaf	R. serpentina	0.74	5.86	2.22	0.52	2.40
	R. verticillata	20.28	15.54	1.85	0.91	3.25
	R. hookeri	0.85	11.59	3.78	0.31	4.86
	R. micrantha	0.64	16.80	6.27	0.43	1.47
	R. tetraphylla	15.79	8.42	0.36	25.19	18.16
	R. vomitoria	2.32	24.24	6.84	1.05	25.23
Herbal formulation (HF)	HF1	0.001	0.03	0.001	nd	0.005
	HF2	0.007	0.24	0.010	nd	0.39
	HF3	0.002	0.27	0.004	nd	0.59
	HF4	0.003	0.05	0.001	nd	0.15

Source: Reproduced from Kumar et al., 2016c with permission from John Wiley & Sons.

in. *R. vomitoria*, whereas it was the least (0.96 mg/g) in *R. verticillata*. In leaves, high contents of ajmaline (20.28 mg/g) and serpentine (25.19 mg/g) were detected in *R. verticillata* and *R. tetraphylla*, respectively, whereas *R. vomitoria* showed the maximum content of yohimbine (24.24 mg/g), ajmalicine (6.84 mg/g) and reserpine (25.23 mg/g). In herbal formulations, HF3 showed highest content of reserpine (0.59 mg) followed by HF2 (0.39 mg/g). Similarly, HF2 showed highest content of ajmaline (0.007 mg/g) and ajmalicine (0.01 mg/g), whereas HF3 showed the maximum yield of yohimbine (0.27 mg/g).

3.5 PRINCIPAL COMPONENT ANALYSIS (PCA)

PCA was applied to statistically establish the correlation and see discrimination among six *Rauvolfia* species on the basis of quantitative analysis data of ajmaline, yohimbine, ajmalicine, serpentine and reserpine. The first two principal components PC1 and PC2 hold 52.89% and 23.60%, respectively, of the total variability in the roots. PCA showed that the roots of *R. micrantha* and *R. hookeri* were close to each other, whereas *R. serpentina*, *R. verticillata* and *R. vomitoria* were falling together in the same quadrants. *R. tetraphylla* was further apart from all other species as shown in Figure 3.3. Similarly, in the leaves, the first two principal components PC1 and PC2 hold 51.33% and 31.52%, respectively, of the total variability.

R. serpentina and *R. verticillata* were falling in near to each other similar to *R. hookeri* and *R. micrantha*, whereas *R. tetraphylla* and *R. vomitoria* were further apart in score plot (Figure 3.3). The quantitative characteristic of *R. tetraphylla* was distinctive in leaves and roots which are placed in the second quadrant of the biplot.

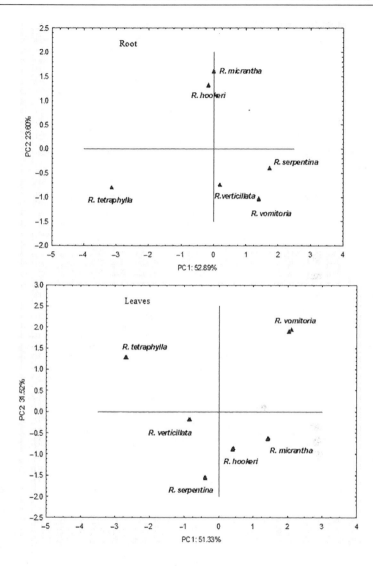

FIGURE 3.3 PC1 vs. PC2 plot showing discrimination among *R. serpentina*, *R. verticillata*, *R. hookeri*, *R. micrantha*, *R. tetraphylla* and *R. vomitoria* roots and leaves on the basis of quantity of ajmaline, yohimbine, ajmalicine, serpentine and reserpine.

3.6 CONCLUSIONS

A rapid and simple LC-MS method was developed for the analysis of six *Rauvolfia* species. The results showed significant qualitative variations among *Rauvolfia* species. These findings will help in the selection of the best suitable plant/part, according to the requirement, and may be used for the authentication and quality control purposes. The marker peaks were identified successfully by PCA discriminating *R. serpentina*, *R. verticillata*, *R. hookeri*, *R. micrantha*, *R. tetraphylla* and *R. vomitoria*. It is evident from this study that PCA effectively served the purpose and all of the six *Rauvolfia* species could be differentiated using this validated method.

References

Abdine, H., M. Abdel-Hady Elsayed, and Yousry M. Elsayed. "Spectrophotometric determination of hydrochlorothiazide and reserpine in combination." *Analyst* 103, no. 1225 (1978): 354–358.

Abubacker, M. N., and S. Vasantha. "Antibacterial activity of ethanolic leaf extracts of Rauwolfia tetraphylla (Apocyanaceae) and its bioactive compound reserpine." *Drug Invention Today* 3, no. 3 (2011): 16–17.

Adu, Francis, John Antwi Apenteng, William Gariba Akanwariwiak, John Henry Sam, David Ntinagyei Mintah, and Edna Beyeman Bortsie. "Antioxidant and in-vitro anthelminthic potentials of methanol extracts of barks and leaves of Voacanga africana and Rauwolfia vomitoria." *African Journal of Microbiology Research* 9, no. 35 (2015): 1984–1988.

Ahmad, Shoaib. "Unani medicine: introduction and present status in India." *The Internet Journal of Alternative Medicine* 6, no. 1 (2008): 8–12.

Ahmedullah, Mohammed, and M. P. Nayar. *Endemic Plants of the Indian Region.* Vol. 1; Calcutta: Botanical Survey of India, 1986.

Alagesaboopathi, C. "An investigation on the antibacterial activity of Rauvolfia tetraphylla dry fruit extracts." *Ethnobotanical Leaflets* 13, (2009): 644–650.

Al-Akraa, Hussen, and Mohamad Kabaweh. "Development and validation of RP-HPLC method for simultaneous determination of clopamide, dihydroergocristine mesylate and reserpine in tablets." *International Journal of Pharmacy and Pharmaceutical Sciences* 7, no. 7 (2015):148–152.

Amer, Mohammed M., and William E. Court. "Leaf alkaloids of Rauwolfia vomitoria." *Phytochemistry* 19, no. 8 (1980): 1833–1836.

Anderson, Melissa A., Timothy Wachs, and Jack D. Henion. "Quantitative ionspray liquid chromatographic/tandem mass spectrometric determination of reserpine in equine plasma." *Journal of Mass Spectrometry* 32, no. 2 (1997): 152–158.

Anonymous. *WHO Monographs on Selected Medicinal Plants.* Vol. I. World Health Organization, Geneva, 1999.

Anonymous. *The Wealth of India: A Dictionary of Indian Raw Materials and Industrial Products.* Publications and Information Directorate; New Delhi: CSIR, 2005: 376–391.

Anonymous. *Ayurvedic Pharmacopoeia of India: Part-1*; New Delhi: Government of India, Department of ISM & H, Published by the Controller of Publications, 2006: 166–167.

Aquaisua, Aquaisua Nyong, Christopher Chiedozie Mbadugha, Enobong Ikpeme Bassey, Moses Bassey Ekong, Theresa Bassey Ekanem, and Monday Isaiah Akpanabiatu. "Effects of rauvolfia vomitoria on the cerebellar histology, body and brain weights of albino wistar rats." *Journal of Experimental and Clinical Anatomy* 16, no. 1 (2017): 41–45.

Arambewela, Lakshmi S. R., and Gamini Madawela. "Alkaloids from Rauvolfia cane-scens." *Pharmaceutical Biology* 39, no. 3 (2001): 239–240.

Azmi, Muhammad Bilal, and Shamim A. Qureshi. "Methanolic root extract of Rauwolfia serpentina benth improves the glycemic, antiatherogenic, and cardioprotective indices in alloxan-induced diabetic mice." *Advances in Pharmacological Sciences* (2012): 376429.

Azmi, Muhammad Bilal, and Shamim A. Qureshi. "Rauwolfia serpentina ameliorates hyperglycemic, haematinic and antioxidant status in alloxan-induced diabetic mice." *Journal of Applied Pharmaceutical Science* 3, no. 7 (2013): 136–141.

Azmi, Muhammad Bilal, and Shamim A. Qureshi. "Rauwolfia serpentina improves altered glucose and lipid homeostasis in fructose-induced type 2 diabetic mice." *Pakistan Journal of Pharmaceutical Sciences* 29, no. 5 (2016): 1619–1624.

Bajpai, Vikas, Awantika Singh, Kamal Ram Arya, Mukesh Srivastava, and Brijesh Kumar. "Rapid screening for the adulterants of Berberis aristata using direct analysis in real-time mass spectrometry and principal component analysis for discrimination." *Food Additives & Contaminants: Part A* 32, no. 6 (2015): 799–807.

Behera, Dipti Ranjan, Rashmi Ranjan Dash, and Sunita Bhatnagar. "Biological evaluation of leaf and fruit extracts of wild snake root (Rauvolfia tetraphylla L.)." *International Journal of Pharmacognosy and Phytochemical Research* 8, no. 7 (2016): 1164–1167.

Beljanski, M., and M. S. Beljanski. "Three alkaloids as selective destroyers of cancer cells in mice." *Oncology* 43, no. 3 (1986): 198–203.

Bhattacharjee, S. K. *Handbook of Medical Plants*; Jaipur: Pointer Publishers, 1998.

Bhishagratna, Kunja Lal, ed. An English translation of *The Sushruta Samhita*: based on original Sanskrit text. Vol. 2; Author, 1911.

Bindu, S., K. B. Rameshkumar, Brijesh Kumar, Awantika Singh, and C. Anilkumar. "Distribution of reserpine in Rauvolfia species from India–HPTLC and LC–MS studies." *Industrial Crops and Products* 62 (2014): 430–436.

Bisong, Sunday, Richard Brown, and Eme Osim. "Comparative effects of Rauwolfia vomitoria and chlorpromazine on social behaviour and pain." *North American Journal of Medical Sciences* 3, no. 1 (2011): 48–54.

Boğa, Mehmet, Murat Bingül. Esra Eroğlu Özkan, and Hasan Şahin. "Chemical and biological perspectives of monoterpene indole alkaloids from Rauwolfia species." In *Studies in Natural Products Chemistry*, Vol. 61; Elsevier, Amsterdam. 2018: 251–299.

Brahmachari, Goutam, Lalan Ch Mandal, Dilip Gorai, Avijit Mondal, Sajal Sarkar, and Sasadhar Majhi. "A new labdane diterpene from Rauvolfia tetraphylla Linn. (Apocynaceae)." *Journal of Chemical Research* 35, no. 12 (2011): 678–680.

Campbell-Tofte, Joan. "Anti-diabetic extract isolated from Rauvolfia vomitoria and Citrus aurantium, and methods of using same." U.S. Patent 7,579,025, issued August 25, 2009.

Chatterjee, A., and A. C. Pakrashi. *The Treatise on Indian Medicinal Plants*. Publication & Information Directorate, Vol. IV; New Delhi: CSIR, 1995: 116–121.

Chatterjee, Asima, and Satyesh Chandra Pakrashi. *The Treatise on Indian Medicinal Plants*; Vol. 1, Publications & Information Directorate, New Delhi, India. 1991.

Chen, Ning, Weixia Li, Shuchao Wu, and Yan Zhu. "Fluorimetric detection of reserpine in mouse serum through online post-column electrochemical derivatization." *Royal Society Open Science* 5, no. 8 (2018): 171948.

Chen, Qinhua, Wenpeng Zhang, Yulin Zhang, Jing Chen, and Zilin Chen. "Identification and quantification of active alkaloids in Catharanthus roseus by liquid chromatography–ion trap mass spectrometry." *Food Chemistry* 139, no. 1–4 (2013): 845–852.

Chopra, R. N., S. L. Nayar, and I. C. Chopra. *Glossary of Indian Medicinal Plants*; New Delhi: National Institute of Science Communication and Information Resources, 2009: 210.

Colpaert, F. C. "Pharmacological characteristics of tremor, rigidity and hypokinesia induced by reserpine in rat." *Neuropharmacology* 26, no. 9 (1987): 1431–1440.

Cronheim, Georg, W. Brown, J. Cawthorne, M. I. Toekes, and J. Ungari. "Pharmacological studies with rescinnamine, a new alkaloid isolated from Rauwolfia serpentina." *Proceedings of the Society for Experimental Biology and Medicine* 86, no. 1 (1954): 120–124.

Cunningham, A. B. *African Medicinal Plants*; Paris: United Nations Educational, Scientific and Cultural Organization, 1993.

Cynthia, N. B., T. K. Dominique, D. M. Tresor, N-N. G. Fredy, K. D. Silue, K.-S. Kouakou-Siransy Gisele, P. H. Joseph, M. Herve, A. A. Antoine, and Y. William. "Antiplasmodial activity and acute oral toxicity of Rauvolfia vomitoria leaves extracts." *International Journal of Current Pharmaceutical Research* 8 (2018): 56–62.

Denis, D. L., Capodice, J. L., Gorrochum, P., Katz, A. E. and Buttyan, R. "Antipostrate cancer activity of a carboline alkaloid enriched extract from *Rauwolfia Vomitoria*." *International Journal of Oncology* 29, no. 5, (2006): 1065–1073.

Deshmukh, Sarika R., Dhanashree S. Ashrit, and Bhausaheb A. Patil. "Extraction and evaluation of indole alkaloids from Rauwolfia serpentina for their antimicrobial and antiproliferative activities." *International Journal of Pharmacy and Pharmaceutical Sciences* 4, no. 5 (2012): 329–334.

Dev, Sukh. "Ancient-modern concordance in Ayurvedic plants: some examples." *Environmental Health Perspectives* 107, no. 10 (1999): 783–789.

Dey, Abhijit, Anuradha Mukherjee, and Madhubrata Chaudhury. "Alkaloids from apocynaceae: origin, pharmacotherapeutic properties, and structure-activity studies." In *Studies in Natural Products Chemistry*; Atta-ur-Rahman, Ed.; Vol. 52; Elsevier, Amsterdam. 2017: 373–488.

Dey, Abhijit, and J. N. De. "Ethnobotanical aspects of Rauvolfia serpentina (L). Benth. ex Kurz. in India, Nepal and Bangladesh." *Journal of Medicinal Plants Research* 5, no. 2 (2011): 144–150.

Diquet, B., L. Doare, and G. Gaudel. "New method for the determination of yohimbine in biological fluids by high-performance liquid chromatography with amperometric detection." *Journal of Chromatography B: Biomedical Sciences and Applications* 311 (1984): 449–455.

Dong, Ruochen, Ping Chen, and Qi Chen. "Inhibition of pancreatic cancer stem cells by Rauwolfia vomitoria extract." *Oncology Reports* 40, no. 6 (2018): 3144–3154.

Duncan, R. J., and C. B. Nash. "Effects of the rauwolfia alkaloids, ajmaline, tetraphyllicine, and serpentine, on myocardial excitability." *Archives internationales de pharmacodynamie et de therapie* 184, no. 2 (1970): 355–361.

D'Urso, Gilda, Luigi d'Aquino, Cosimo Pizza, and Paola Montoro. "Integrated mass spectrometric and multivariate data analysis approaches for the discrimination of organic and conventional strawberry (Fragaria ananassa Duch.) crops." *Food Research International* 77 (2015): 264–272.

Dutta, Shubhra, Anindya Roy Chowdhury, S. K. Srivastava, Ilora Ghosh, and Kasturi Datta. "Evidence for Serpentine as a novel antioxidant by a redox sensitive HABP1 overexpressing cell line by inhibiting its nuclear translocation of NF-κB." *Free Radical Research* 45, no. 11–12 (2011): 1279–1288.

Dwivedi, Binit Kumar, Manoj Kumar, Bhopal Singh Arya, Echur Natrajan Sundaram, and Raj K. Manchanda. "Comparative standardization study for determination of reserpine in Rauwolfia serpentina homoeopathic mother tinctures manufactured by different pharmaceutical industries using HPTLC as a check for quality control." *Indian Journal of Research in Homoeopathy* 11, (2017): 109–117.

Ebuehi, Osaretin A. T., Iroghama Asoro, Miriam N. Igwo-Ezikpe, Ngozi O. Imaga, Ochuko L. Erukainure, Ridwan A. Lawal, Sunday Adenekan, Uba Duncan, and Chijoke Micah. "Comparative toxicity studies of Rauwolfia vomitoria leaf and root extracts in Wistar rats." *International Journal of Biological and Medical Research* 9, no. 2 (2018): 6357–6362.

Ekor, Martins. "The growing use of herbal medicines: issues relating to adverse reactions and challenges in monitoring safety." *Frontiers in Pharmacology* 4 (2014): 177.

El-Din, Mohie M. K. Sharaf, Mohamed W. I. Nassar, Khalid A. M. Attia, Maha A. El Demellawy, and Mohamed M. Y. Kaddah. "Validated liquid chromatography–tandem mass spectrometry method for simultaneous determination of clopamide, reserpine and dihydroergotoxine: Application to pharmacokinetics in human plasma." *Journal of Pharmaceutical and Biomedical Analysis* 125 (2016): 236–244.

Elisabetsky, E., and L. Costa-Campos. "The alkaloid alstonine: a review of its pharmacological properties." *Evidence-Based Complementary and Alternative Medicine* 3, no. 1 (2006): 39–48.

Eluwa, M., N. Idumesaro, M., Ekong, A. Akpantah, and T. Ekanem. "Effect of aqueous extract of rauwolfia vomitoria root bark on the cytoarchitecture of the cerebellum and neurobehaviour of adult male wistar rats." *The Internet Journal of Alternative Medicine* 6, no. 2 (2008): 1–7.

Elyushnichenko, V. E., Yakimov, S. A., Tuzova, T., Syagailo, Y. V., Kuzovkina, I. N., Wulfson, A. N., Miroshnikova A. I. "Determination of indole alkaloids from *R. serpentina* and *R. vomitoria* by high-performance liquid chromatography and high-performance thin-layer chromatography." *Journal of Chromatography A* 704 (1995): 357–362.

Erasto, Paul, Antoinette Lubschagne, Zakaria H. Mbwambo, Ramadhani S. O. Nondo, and Namrita Lall. "Antimycobacterial, antioxidant activity and toxicity of extracts from the roots of Rauvolfia vomitoria and R. caffra." *Spatula DD* 1 no. 2 (2011): 73–80.

Ezeigbo, I. I., M. I. Ezeja, K. G. Madubuike, D. C. Ifenkwe, I. A. Ukweni, N. E. Udeh, and S. C. Akomas. "Antidiarrhoeal activity of leaf methanolic extract of Rauwolfia serpentina." *Asian Pacific Journal of Tropical Biomedicine* 2, no. 6 (2012): 430–432.

Fapojuwomi, O. A., and I. O. Asinwa. "Assessment of medicinal values of Rauvolfia vomitoria (Afzel) in Ibadan Municipality." *Greener Journal of Medical Sciences* 3, no. 2 (2013): 37–41.

Farnsworth, Norman R., and D. D. Soejarto. "Global importance of medicinal plants." In: *The Conservation of Medicinal Plants*; Akerele O., Heywood V. and Synge H., Eds.; Cambridge University Press, Cambridge, UK, Vol. 26; 1991: 25–51.

Farooqi, A. A., and B. S. Sreeramu. *Cultivation of Medicinal and Aromatic Plants*; Hyderabad: University Press, 2004.

Forni, G. P. "Gas chromatographic determination of ajmaline in the dark of the root of Rauvolfa vomitoria." *Journal of Chromatography A* 176, no. 1 (1979): 129–133.

Friedrich, O., F. v Wegner, M. Wink, and R. H. A. Fink. "NA+- and K+-channels as molecular targets of the alkaloid ajmaline in skeletal muscle fibres." *British Journal of Pharmacology* 151, no. 1 (2007): 63–74.

Gadvi, R., M. N. Reddy, and M. Nivserkar. "Antihypertensive efficacy of *Rouwolfia* tetraphylla–root of the plant on uninephrectomized DOCA–salt hypertensive rats." *International Journal of Pharmaceutical Sciences and Research* 9, no. 3 (2018): 1251–1255.

Ganapaty, S., Thomas P. STEVE, Ramana K. Venkata, and V. Neeharika. "A review of phytochemical studies of Rauwolfia species." *Indian Drugs* 38, no. 12 (2001): 601–612.

Ganugapati, Jayasree, Aashish Baldwa, and Sarfaraz Lalani. "Docking studies of Rauwolfia serpentina alkaloids as insulin receptor activators." *International Journal of Computers and Applications* 43, no. 14 (2012): 32–37.

Gao, Yuan, Ai-Lin Yu, Gen-Tao Li, Ping Hai, Yan Li, Ji-Kai Liu, and Fei Wang. "Hexacyclic monoterpenoid indole alkaloids from Rauvolfia verticillata." *Fitoterapia* 107 (2015a): 44–48.

Gao, Yuan, Dong-Sheng Zhou, Ling-Mei Kong, Ping Hai, Yan Li, Fei Wang, and Ji-Kai Liu. "Rauvotetraphyllines AE, new indole alkaloids from Rauvolfia tetraphylla." *Natural Products and Bioprospecting* 2, no. 2 (2012): 65–69.

Gao, Yuan, Dong-Sheng Zhou, Ping Hai, Yan Li, and Fei Wang. "Hybrid monoterpenoid indole alkaloids obtained as artifacts from Rauvolfia tetraphylla." *Natural Products and Bioprospecting* 5, no. 5 (2015b): 247–253.

Government of India Planning Commission. *Report of the Task Force on Conservation and Sustainable Use of Medicinal Plants*; New Delhi: Government of India Planning Commission; 2000, http://planningcommission.gov.in/aboutus/task-force/tsk_medi.pdf. Accessed January 9, 2015.

Gupta, Ajay K., Havagiray Chitme, Sujata K. Dass, and Neelam Misra. "Hepatoprotective activity of Rauwolfia serpentina rhizome in paracetamol intoxicated rats." *Journal of Pharmacological and Toxicological* 1, no. 1 (2006a): 82–88.

Gupta, Jaya, and Amit Gupta. "Isolation and extraction of flavonoid from the leaves of Rauwolfia serpentina and evaluation of DPPH-scavenging antioxidant potential." *Oriental Journal of Chemistry* 31, no. Special Issue 1 (2015a): 231–235.

Gupta, Jaya, and Amit Gupta. "Isolation and identification of flavonoid rutin from Rauwolfia serpentina." *International Journal of Chemical Studies* 3, no. 2 (2015b): 113–115.

Gupta, Jaya, Amit Gupta, and A. K. Gupta. "Extraction and identification of flavonoid natural antioxidant in the leaves of Rauwolfia serpentina." *IJCS* 3, no. 1 (2015): 35–37.

Gupta, Madan, Alpana Srivastava, Arvind Tripathi, Himanshu Misra, and Ram Verma. "Use of HPTLC, HPLC, and densitometry for qualitative separation of indole alkaloids from Rauvolfia serpentina roots." *JPC-Journal of Planar Chromatography-Modern TLC* 19, no. 110 (2006b): 282–287.

Gupta, Shikha, Vinay Kumar Khanna, Anupam Maurya, Dnyaneshwar Umrao Bawankule, Rajendra Kumar Shukla, Anirban Pal, and Santosh Kumar Srivastava. "Bioactivity guided isolation of antipsychotic constituents from the leaves of Rauwolfia tetraphylla L." *Fitoterapia* 83, no. 6 (2012): 1092–1099.

Hai-Bo, Liu, Peng Yong, Huang Lu-qi, Xu Jun, and Xiao Pei-Gen. "Mechanism of selective inhibition of yohimbine and its derivatives in adrenoceptor α2 subtypes." *Journal of Chemistry* (2013): 783058.

Hariga, M. "Preliminary trial of therapeutic use of reserpiline, a hypotensive extract of Rauwolfia vomitoria." *Bruxelles Medical* 39, no. 13 (1959): 451–455.

Hong, Bo, Weiming Cheng, Jian Wu, and Chunjie Zhao. "Screening and identification of many of the compounds present in Rauvolfia verticillata by use of high-pressure LC and quadrupole TOF MS." *Chromatographia* 72, no. 9–10 (2010): 841–847.

Hong, Bo, Wen-Jing Li, and Chun-Jie Zhao. "Chemical constituents of *Rauvolfia verticillata*." *Yao xue xue bao= Acta pharmaceutica Sinica* 47, no. 6 (2012a): 764–768.

Hong, Bo, Wen-Jing Li, Ai-Hua Song, and Chun-Jie Zhao. "Determination of indole alkaloids and highly volatile compounds in Rauvolfia verticillata by HPLC–UV and GC–MS." *Journal of Chromatographic Science* 51, no. 10 (2012b): 926–930.

Hong, B., Li, W., Song, A. and Zhao, C. "Determination of indole alkaloids and highly volatile compounds in *Rauvolfia verticillata* by HPLC–UV and GC–MS." *Journal of Chromatographic Science* 51 (2013): 926–930.

Ian, J. *Principal Component Analysis*; Hoboken: John Wiley & Sons, 2002.

Iqbal, Amjad Ali M., Firoz A. Kalam Khan, Imtiyaz Ansari, Altamash Quraishi, and Mohib Khan. "Ethno-phyto-pharmacological overview on Rauwolfia densiflora (Wall) Benth. ex Hook. f." *International Journal of Pharmaceutical and Phytopharmacological Research* 2 (2013a): 372–376.

Iqbal, Amjad Ali M., Firoz A. Kalam Khan, and Mohib Khan. "Ethno-Phyto-Pharmacological Overview on Rauwolfia tetraphylla L." *International Journal of Pharmaceutical and Phytopharmacological Research* 2, no. 4 (2013b): 247–251.

Iqbal, Muzaffar, Aftab Alam, Tanveer A. Wani, and Nasr Y. Khalil. "Simultaneous determination of reserpine, rescinnamine, and yohimbine in human plasma by ultraperformance liquid chromatography tandem mass spectrometry." *Journal of Analytical Methods in Chemistry* (2013c): 940861 11 pages.

Itoh, Atsuko, Tomoko Kumashiro, Machiko Yamaguchi, Naotaka Nagakura, Yoshiyuki Mizushina, Toyoyuki Nishi, and Takao Tanahashi. "Indole alkaloids and other constituents of Rauwolfia serpentina." *Journal of Natural Products* 68, no. 6 (2005): 848–852.

Ivanova, Stanislava, Kalin Ivanov, Stanislav Gueorgiev, and Elina Petkova. "UHPLC/ MS detection of yohimbine in food supplements." *Biomedical Research* 28, no. 7 (2017): 3281–3285.

Iwu, M. M. "Stem bark alkaloids of Rauwolfia vomitoria." *Planta medica* 45, no. 6 (1982): 105–111.

Iwu, Maurice M. "Root alkaloids of Rauwolfia vomitoria Afz." *Planta medica* 32, no. 5 (1977): 88–94.

Jakaria, M. D., S. M. Tareq, M. Ibrahim, and S. Bokhtearuddin. "Rauvolfia tetraphylla L.(Apocynaceae): A Pharmacognostical, Phytochemical and Pharmacological Review." *Journal of Chemical and Pharmaceutical Research* 8, no. 12 (2016): 114–120.

Jeong, Won Tae, and Heung Bin Lim. "A UPLC-ESI-Q-TOF method for rapid and reliable identification and quantification of major indole alkaloids in *Catharanthus roseus*." *Journal of Chromatography B* 1080 (2018): 27–36.

Kamboj, V. P. "Herbal medicine." *Current Science* 78 (2000): 35–51.

Kannaiyan, S. "Diversity, sustainable use and conservation of medicinal plants." Inaugural address to the *International Seminar on Medicinal Plants and Herbal Products*, Tirupati, March 7–9, 2008. http://ismphpabstracts.blogspot.in/. Accessed August 20, 2014.

Kanyal, Neema. "Role of Rauwolfia serpentina in stroke induced experimental dementia." *Indian Journal of Pharmaceutical and Biological Research* 4, no. 1 (2016): 19–30.

Karioti, A., E. Giocaliere, C. Guccione, G. Pieraccini, E. Gallo, A. Vannacci, and A. R. Bilia. "Combined HPLC-DAD–MS, HPLC–MSn and NMR spectroscopy for quality control of plant extracts: the case of a commercial blend sold as dietary supplement." *Journal of Pharmaceutical and Biomedical Analysis* 88 (2014): 7–15.

Klohs, M. W., M. D. Draper, and F. Keller. "Alkaloids of Rauwolfia serpentina Benth. iii. 1 rescinnamine, a new hypotensive and sedative principle." *Journal of the American Chemical Society* 76, no. 10 (1954): 2843–2843.

Kokate, C. K., A. P. Purohit, and S. B. Gokhale. *Pharmacognosy*, 24th ed.; Pune: Nirali Prakashan, 2003: 466–470.

Köppel, C., A. Wagemann, and F. Martens. "Pharmacokinetics and antiarrhythmic efficacy of intravenous ajmaline in ventricular arrhythmia of acute onset." *European Journal of Drug Metabolism and Pharmacokinetics* 14, no. 2 (1989): 161–167.

Koul, Pamposh Mohan, and Basanagouda Sanganagouda Janagoudar. "Phytochemical analysis of roots among various genotypes across different age groups in *Rauwolfia Serpentina* (L.) benth by high performance liquid chromatography." *International Journal of pharma and Bio Science* 8, no. 1 (2017): 35–40.

Kumar, A., M. K. Bhardwaj, A. K. Upadhyay, A. Tiwari, and B. D. Ohdar. "Quantitative determination of Yohimbine alkaloid in the different part of the Rauvolfia tetraphylla." *Journal of Chemical and Pharmaceutical Research* 3, no. 2 (2011): 907–910.

Kumar, J. U. S., M. J. K. Chaitanya, A. J. Semotiuk, and V. Krishna. "Indigenous knowledge of medicinal plants used by ethnic communities of South India." *Journal of Ethnopharmacology* 18 (2019): 1–112.

Kumar, Sunil, Awantika Singh, Vikas Bajpai, and Brijesh Kumar. "Identification, characterization and distribution of monoterpene indole alkaloids in *Rauwolfia* species by Orbitrap Velos Pro mass spectrometer." *Journal of Pharmaceutical and Biomedical Analysis* 118 (2016a): 183–194.

Kumar, Sunil, Awantika Singh, Vikas Bajpai, Mukesh Srivastava, Bhim Pratap Singh, and Brijesh Kumar. "Structural characterization of monoterpene indole alkaloids in ethanolic extracts of Rauwolfia species by liquid chromatography with quadrupole time-of-flight mass spectrometry." *Journal of Pharmaceutical Analysis* 6, no. 6 (2016b): 363–373.

Kumar, Sunil, Awantika Singh, Vikas Bajpai, Mukesh Srivastava, Bhim Pratap Singh, Sanjeev Ojha, and Brijesh Kumar. "Simultaneous determination of bioactive monoterpene indole alkaloids in ethanolic extract of seven Rauvolfia species using UHPLC with hybrid triple quadrupole linear ion trap mass spectrometry." *Phytochemical Analysis* 27, no. 5 (2016c): 296–303.

Kumar, Sunil, Vikas Bajpai, Awantika Singh, S. Bindu, Mukesh Srivastava, K. B. Rameshkumar, and Brijesh Kumar. "Rapid fingerprinting of Rauwolfia species using direct analysis in real time mass spectrometry combined with principal component analysis for their discrimination." *Analytical Methods* 7, no. 14 (2015): 6021–6026.

Kumara, P. Mohana, R. Uma Shaanker, and T. Pradeep. "UPLC and ESI-MS analysis of metabolites of Rauvolfia tetraphylla L. and their spatial localization using desorption electrospray ionization (DESI) mass spectrometric imaging." *Phytochemistry* 159 (2019): 20–29.

Kumari, R., B. Rathi, A. Rani, and Sonal Bhatnagar. "Rauvolfia serpentina L. Benth. ex Kurz.: phytochemical, pharmacological and therapeutic aspects." *International Journal of Pharmaceutical Sciences Review and Research* 23, no. 2 (2013): 348–355.

Kunakh, V. A. "Somaclonal variation in Rauwolfia." In *Somaclonal Variation in Crop Improvement II*; Berlin and Heidelberg: Springer, 1996: 315–332.

Kurian, A., M. K. Arjunan, B. Thomas, D. P. Sebastian, and S. George. "Effect of extraction solvents on phytochemicals of various parts of R. hookeri S. R. Srini. and Chitra- An endemic medicinal plant of Western Ghats." *European Journal of Biomedical and Pharmaceutical Sciences* 4 (2017): 517–523.

Lambert, Geoffrey A., William J. Lang, Eitan Friedman, Emanuel Meller, and Samuel Gershon. "Pharmacological and biochemical properties of isomeric yohimbine alkaloids." *European Journal of Pharmacology* 49, no. 1 (1978): 39–48.

Li, Hang, Junting He, Qin Liu, Zhaohui Huo, Si Liang, Yong Liang, and Yoichiro Ito. "Simultaneous determination of hydrochlorothiazide and reserpine in human urine by LC with a simple pre-treatment." *Chromatographia* 73, no. 1–2 (2011): 171–175.

Li, P.-T., A. J. M. Leeuwenberg, and D. J. Middleton. "Apocynaceae." In *Flora of China, Vol. 16, Gentianaceae through Boraginaceae*; Flora of China Editorial Committee; Ed.; St. Louis: Science Press; 1995: 143–188; 479 pp.

Lin, Mao, De-Quan Yu, Xin Liu, F. Fu, Q. Zheng, C. He, G. Bao, and C. Xu. "Chemical studies on the quaternary alkaloids of Rauvolfia verticillata (Lour.) Baill. F. ruberocarpa HT Chang. mss." *Acta Pharmaceutica Sinica* 20 (1985): 198–202.

Linck, Viviane M., Ana P. Herrmann, Ângelo L. Piato, Bernardo C. Detanico, Micheli Figueiró, Jorge Flório, Maurice M. Iwu, Christopher O. Okunji, Mirna B. Leal, and Elaine Elisabetsky. "Alstonine as an antipsychotic: effects on brain amines and metabolic changes." *Evidence-Based Complementary and Alternative Medicine* 418597 (2011) doi:10.1093/ecam/nep002.

Linnaeus, C. Species Plantarum. Laurentius Salvius, Stockholm, 1753.

Liu, J., Y. Liu, Y. J. Pan, Y. G. Zu, and Z. H. Tang. "Determination of alkaloids in Catharanthus roseus and Vinca minor by high-performance liquid chromatography–tandem mass spectrometry." *Analytical Letters* 49 (2015): 1143–1153.

Lobay, Douglas. "Rauwolfia in the treatment of hypertension." *Integrative Medicine: A Clinician's Journal* 14, no. 3 (2015): 40–46.

Lohani, Hema, Harish Chandra Andola, Ujjwal Bhandari, and Nirpendra Chauhan. "HPTLC method validation of reserpine in Rauwolfia serpentine – A high value medicinal plant." *Researcher* 3 (2011): 34–37.

Lovati, M., F. Peterlongo, T. Ruffilli, and G. F. Zini. "Two new indole alkaloids from Rauwolfia vomitoria." *Fitoterapia* 67, no. 5 (1996): 422–426.

Mabberley, David J. *Mabberley's Plant-Book: A Portable Dictionary of Plants, their Classification and Uses*, 4th ed.; Cambridge: Cambridge University Press, 2017.

Mahalakshmi, S. N., Achala, H. G., Ramyashree, K. R. and Prashith Kekuda, T. R. "Rauvolfia tetraphylla L. (Apocynaceae) – a comprehensive review on its ethnobotanical uses, phytochemistry and pharmacological activities." *International Journal of Pharmacy and Biological Sciences-IJPBSTM* 9, no. 2 (2019): 664–682.

Mahata, Manjula, Sushil K. Mahata, Robert J. Parmer, and Daniel T. O'Connor. "Vesicular monoamine transport inhibitors: novel action at calcium channels to prevent catecholamine secretion." *Hypertension* 28, no. 3 (1996): 414–420.

Malik, A., and S. Siddiqui. "The subsidiary alkaloids of Rauwolfia vomitoria Afzuelia." Pakistan *Journal of Scientific and Industrial Research* 22 (1979): 121–123.

McQueen, E. G., A. E. Doyle, and F. H. Smirk. "Mechanism of hypotensive action of reserpine, an alkaloid of Rauwolfia serpentina." *Nature* 174, no. 4439 (1954): 1015–1015.

Momoh, J., O. O. Aina, S. M. Akoro, O. Ajibaye, and H. I. Okoh. "In Vivo Anti-Plasmodial Activity and the Effect of Ethanolic Leaf Extract of Rauvolfia Vomitoria on hematological and Lipid Parameters in Swiss Mice Infected with Plasmodium Berghei NK 65." *Nigerian Journal of Parasitology* 35, no. 1&2 (2014): 109–116.

Monachino, Joseph. "Rauvolfia serpentina—its history, botany and medical use." *Economic Botany* 8, no. 4 (1954): 349–365.

Mukerji, B. "Concluding remarks on *Rauvolfias* in modern therapy." *Indian Journal of Pharmacy* 18 (1956): 433–440.

Mukerji, B. "India's wonder drug plant-Rauwolfia serpentina: birth of a new drug from an old Indian medicinal plant." *Proceedings of the National Academy of Sciences, India Section B: Biological Sciences* 42 (1976): 1–11.

Müller, J. M., E. Schlittler, and H. J. Bein. "Reserpin, der sedative Wirkstoff ausRauwolfia serpentina Benth." *Cellular and Molecular Life Sciences* 8, no. 9 (1952): 338–338.

Muller, Mason. "Rauwolfia serpentina (Sarpagandha) Medicinal herb uses and pictures" https://www.homeremediess.com/medicinal-herb-sarpagandha-uses/ www.homeremediess.com, 2015.

N'doua, Léatitia Akouah Richmonde, Kouakou Jean Claude Abo, Serge Aoussi, Léandre Kouakou Kouakou, and Etienne Ehouan Ehile. "Aqueous extract of Rauvolfia vomitoria Afzel (Apocynaceae) roots effect on blood glucose level of normoglycemic and hyperglycemic rats." *American Scientific Research Journal for Engineering, Technology, and Sciences (ASRJETS)* 20, no. 1 (2016): 66–77.

Nair, Vadakkemuriyil Divya, Rajaram Panneerselvam, and Ragupathi Gopi. "Studies on methanolic extract of Rauvolfia species from Southern Western Ghats of India–In vitro antioxidant properties, characterisation of nutrients and phytochemicals." *Industrial Crops and Products* 39 (2012): 17–25.

Nair, Vadakkemuriyil Divya, Rajaram Panneerselvam, and Ragupathi Gopi. Flavonoid as chemotaxonomic markers in endemic/endangered species of Rauvolfia from Southern Western Ghats of India: A preliminary study." *Plant Biosystems— An International Journal Dealing with all Aspects of Plant Biology* 147, no. 3 (2013): 704–712.

Nandhini, V. S., and G. Viji Stella Bai. "In-vitro phytopharmacological effect and cardio protective activity of Rauvolfia tetraphylla L." *South Indian Journal of Biological Sciences* 1, no. 2 (2015): 97–102.

Negi, J. S., V. K. Bisht, A. K. Bhandari, D. S. Bisht, P. Singh, and N. Singh. "Quantification of reserpine content and antibacterial activity of Rauvolfia serpentina (L.) Benth. ex Kurz." *African Journal of Microbiology Research* 8, no. 2 (2014): 162–166.

Ng, F. S. P. *Tropical Horticulture and Gardening*; Kuala Lumpur: Clearwater Publications, 2006.

Obreshkova, D., and D. Tsvetkova. "Validation of HPLC method with UV-detection for determination of yohimbine containing products." *Pharmacia* 63 (2016): 3–9.

Olatokunboh, Amole Olufemi, Yemitan Omoniyi Kayode, and Oshikoya Kazeem Adeola. "Anticonvulsant activity of Rauvolfia vomitoria (Afzel)." *African Journal of Pharmacy and Pharmacology* 3, no. 6 (2009): 319–322.

Omotayo, Felix Oluwafemi, and Temitope Israel Borokini. "Comparative phytochemical and ethnomedicinal survey of selected medicinal plants in Nigeria." *Scientific Research and Essays* 7, no. 9 (2012): 989–999.

Owk, A. K., and Mutyala N. Lagudu. "In-vitro Antimicrobial Activity of Roots of Rauwolfia serpentina L. Benth Kurz." *Notulae Scientia Biologicae* 8, no. 3 (2016): 312–316.

Pandey, Devendra Kumar, and Abhijit Dey. "A validated and densitometric HPTLC method for the simultaneous quantification of reserpine and ajmalicine in Rauvolfia serpentina and Rauvolfia tetraphylla." *Revista Brasileira de Farmacognosia* 26, no. 5 (2016): 553–557.

Patel, M. B., J. Poisson, J. L. Pousset, and J. M. Rowson. "Alkaloids of the leaves of Rauwolfia vomitoria Afz." *Journal of Pharmacy and Pharmacology* 16, no. S1 (1964): 163T–165T.

Patel, U. D., H. B. Patel, and B. B. Javia. "Antimicrobial potency of Rauvolfia tetraphylla and Jatropha curcas." *Wayamba Journal of Animal Science* 5 (2013): 723–728.

Pathak, D. K., P. Pandita, M. Bisht, H. S. Choudhary, and R. Yadav. "Structure determination of Rauwolfia Serpentina Benth water soluble seed polysaccharide by methylation studies." *International Journal of Chemical Engineering and Applications* 4 (2012): 63–70.

Pathania, Shivalika, Sai Mukund Ramakrishnan, Vinay Randhawa, and Ganesh Bagler. "SerpentinaDB: a database of plant-derived molecules of Rauvolfia serpentina." *BMC Complementary and Alternative Medicine* 15, no. 1 (2015): 262.

Pathania, Shivalika, Vinay Randhawa, and Ganesh Bagler. "Prospecting for novel plant-derived molecules of Rauvolfia serpentina as inhibitors of Aldose Reductase, a potent drug target for diabetes and its complications." *PLoS ONE* 8, no. 4 (2013): e61327.

Plumier, Charles. *Nova Plantarum Americanarum Genera,* Parisiis: apud Joannem Boudot, 1703.

Qureshi, S. A., A. Nawaz, S. K. Udani, and B. Azmi. "Hypoglycaemic and hypolipidemic activities of Rauvolfia serpentina in alloxan-induced diabetic rats." *International Journal of Pharmacology* 5, no. 5 (2009): 323–326.

Qureshi, Shamim A., and Shamsa K. Udani. "Hypolipidaemic activity of Rauwolfia serpentina Benth." *Pakistan Journal of Nutrition* 8, no. 7 (2009): 1103–1106.

Rajasree, P. H., Ranjith Singh, and C. Sanskar. "Anti-venom activity of ethanolic extract of Rauwolfia serpentine against Naja naja (Cobra) venom." *International Journal of Drug Discovery and Herbal Research* 3 (2013): 521–524.

Ramos, Alexander E. Fox, Pierre Le Pogam, Charlotte Fox Alcover, Elvis Otogo N'Nang, Gaëla Cauchie, Hazrina Hazni, Khalijah Awang, et al. "Collected mass spectrometry data on monoterpene indole alkaloids from natural product chemistry research." *Scientific Data* 6, no. 1 (2019): 15.

Rao, B. Ganga, P. Umamaheswara Rao, E. Sambasiva Rao, and T. Mallikarjuna Rao. "Evaluation of in-vitro antibacterial activity and anti-inflammatory activity for different extracts of Rauvolfia tetraphylla L. root bark." *Asian Pacific Journal of Tropical Biomedicine* 2, no. 10 (2012): 818–821.

Roberts, Margaret F (University of London)., Michael Wink (University of Heidelberg) Ed. *Alkaloids: Biochemistry, Ecology, and Medicinal Applications*; Springer Science & Business Media, Plenum Press, New York, (1998).

Rohela, Gulab Khan, Prasad Bylla, Rajender Korra, and Christopher Reuben. "Phytochemical screening and antimicrobial activity of leaf, stem, root and their callus extracts in Rauwolfia tetraphylla." *International Journal of Agriculture & Biology* 18, no. 3 (2016): 521–528.

Rolf, Sascha, Hans-Jürgen Bruns, Thomas Wichter, Paulus Kirchhof, Michael Ribbing, Kristina Wasmer, Matthias Paul, Günter Breithardt, Wilhelm Haverkamp, and Lars Eckardt. "The ajmaline challenge in Brugada syndrome: diagnostic impact, safety, and recommended protocol." *European Heart Journal* 24, no. 12 (2003): 1104–1112.

Ronchetti, F., G. Russo, E. Bombardelli, and A. Bonati. "A new alkaloid from rauwolfia vomitoria." *Phytochemistry* 10, no. 6 (1971): 1385–1388.

Rukachaisirikul, Thitima, Suwadee Chokchaisiri, Parichat Suebsakwong, Apichart Suksamrarn, and Chainarong Tocharus. "A new Ajmaline-type alkaloid from the roots of *Rauvolfia* serpentina." *Natural Product Communications* 12, no. 4 (2017): 495–498.

Sabri, Nawal N., and William E. Court. "Stem alkaloids of Rauwolfia vomitoria." *Phytochemistry* 17, no. 11 (1978): 2023–2026.

Sachdev, K. S., Ranita Aiman, and M. V. Rajapurkar. "Antihistaminase activity of serpentine." *British Journal of Pharmacology and Chemotherapy* 16, no. 2 (1961): 146–152.

Sagi, S., B. Avula, Y. H. Wang, and I. A. Khan. "Quantification of alkaloids from roots of Rauwolfia species and dietary supplements using UHPLC-UV." *Planta Medica* 81, no. 5 (2015): PA22.

Sagi, Satyanarayanaraju, Bharathi Avula, Yan-Hong Wang, and Ikhlas A. Khan. "Quantification and characterization of alkaloids from roots of Rauwolfia serpentina using ultra-high performance liquid chromatography-photo diode array-mass spectrometry." *Analytical and Bioanalytical Chemistry* 408, no. 1 (2016): 177–190.

Sahu, B. N. *Taxonomy of Indian Species, Rauvolfia serpentina*; New Delhi: Today and Tomorrows Printers and Publishers, 1979: 70–71.

Sastri, K. *Charaka samhita of agnivesa of cakrapanidatta. part-ii. Chikitsasthanam*; Varanasi: chaukhambha sankrit sansthan, 2006: 582.

Schneider, Jurg A., A. J. Plummer, A. E. Earl, W. E. Barrett, R. Rinehart, and R. C. Dibble. "Pharmacological studies with deserpidine, a new alkaloid from Rauwolfia canescens." *Journal of Pharmacology and Experimental Therapeutics* 114, no. 1 (1955): 10–13.

Sen, Saikat, and Raja Chakraborty. "Toward the integration and advancement of herbal medicine: a focus on traditional Indian medicine." *Botanics: Targets and Therapy* 5, no. 33 (2015): e44.

Settimj, Guido, Luciano Di Simone, and Maria Rosaria Del Giudice. "A new gas chromatographic method for the estimation of reserpine and rescinnamine." *Journal of Chromatography A* 116, no. 2 (1976): 263–270.

Sharma, Neelam, and K. P. S. Chandel. "Low-temperature storage of Rauvolfia serpentina Benth. ex Kurz.: an endangered, endemic medicinal plant." *Plant Cell Reports* 11, no. 4 (1992): 200–203.

Siddiqui, Salimuzzaman, S. Imtiaz Haider, and S. Salman Ahmad. "Indobine—a New Alkaloid from Rauwolfia serpentina Benth." *Zeitschrift für Naturforschung B* 42, no. 6 (1987): 783–784.

Singh, Aaditya, Shalini Tripathi, and P. Singh. "Anticonvulsant activity of rauwolfia tetraphylla leaf extract in swiss albino mice." *Asian Journal of Pharmaceutical and Clinical Research* 12, no. 2 (2019): 377–380.

Somers, K. "Notes on Rauwolfia and ancient medical writings of India." *Medical History* 2, no. 2 (1958): 87–91.

Sreekumar, S., N. C. Nisha, C. K. Biju, and P. N. Krishnan. "Identification of potential lead compounds against cobra venom in Rauvolfia serpentina (L.) Benth. Ex kurz through molecular docking." *International Journal of Pharmaceutical Research and Development* 6 (2014): 32–43.

Srivastava, A., A. K. Tripathi, R. Pandey, R. K. Verma, and M. M. Gupta. "Quantitative determination of reserpine, ajmaline, and ajmalicine in Rauvolfia serpentina by reversed-phase high-performance liquid chromatography." *Journal of Chromatographic Science* 44, no. 9 (2006): 557–560.

Stansbury, Jill, Paul Saunders, David Winston, and Eugene R. Zampieron. "Reversing Hypertension with Rauwolfia, Viscum and Piscidia." *Journal of Restorative Medicine* 1, no. 1 (2012): 96–101.

Stöckigt, Joachim, Matthias Unger, Detlef Stöckigt, and Detlev Belder. "Analysis of alkaloids by capillary electrophoresis and capillary electrophoresis-electrospray mass spectrometry." *Alkaloids: Chemical and Biological Perspectives*; S. William Pelletier, Ed., Elsevier, Amsterdam.12 (1998): 289–341.

Sun, Jianghao, Andrew Baker, and Pei Chen. "Profiling the indole alkaloids in yohimbe bark with ultra-performance liquid chromatography coupled with ion mobility quadrupole time-of-flight mass spectrometry." *Rapid Communications in Mass Spectrometry* 25, no. 18 (2011): 2591–2602.

Suresh, K., S. Saravana Babu, and R. Harisaranraj. "Studies on in vitro antimicrobial activity of ethanol extract of Rauvolfia tetraphylla." *Ethnobotanical Leaflets* 12, (2008): 586–590.

Tekwu, Emmanuel Mouafo, Kwabena Mante Bosompem, William Kofi Anyan, Regina Appiah-Opong, Kofi Baffour-Awuah Owusu, Mabel Deladem Tettey, Felicia Amanfo Kissi, Alfred Ampomah Appiah, Veronique Penlap Beng, and Alexander Kwadwo Nyarko. "In vitro assessment of anthelmintic activities of Rauwolfia vomitoria (Apocynaceae) stem bark and roots against parasitic stages of Schistosoma mansoni and cytotoxic study." *Journal of Parasitology Research* (2017): 2583969 11 pages.

Thinakaran, T., A. Rajendran, and V. Sivakumari. "Pharmacognostical, phytochemical and pharmacological studies in Rauvolfia tetraphylla L." *Asian Journal of Environmental Science* 4, no. 1 (2009): 81–85.

Uhlig, Silvio, Wolfgang Egge-Jacobsen, Trude Vrålstad, and Christopher O. Miles. "Indole–diterpenoid profiles of Claviceps paspali and Claviceps purpurea from high-resolution Fourier transform Orbitrap mass spectrometry." *Rapid Communications in Mass Spectrometry* 28, no. 14 (2014): 1621–1634.

Vakil, Rustom Jal. "Rauwolfia serpentina in the treatment of high blood pressure: a review of the literature." *Circulation* 12, no. 2 (1955): 220–229.

Varier, M. R. Raghava. "Origins and growth of āyurvedic knowledge." *Indian Journal of History of Science* 51 (2016): 40–47.

Verma, Ram Kishore, Shikha Gupta, Madan Mohan Gupta, and Santosh Kumar Srivastava. "A simple isocratic HPLC method for the simultaneous determination of antipsychotic indole alkaloids in Rauwolfia tetraphylla." *Journal of Liquid Chromatography & Related Technologies* 35, no. 7 (2012): 937–950.

Vinay, K. N., V. Venkata Lakshmi, N. D. Satyanarayan, and G. R. Anantacharya. "Antioxidant Activity of Leaf And Fruit Extracts Of Rauwolfia Tetraphylla Linn." *International Journal of Pharmaceutical Sciences and Research* 7, no. 4 (2016): 1705–1709.

Wang, Pei, Lian Li, Hailong Yang, Shijuan Cheng, Yingzi Zeng, Lei Nie, and Hengchang Zang. "Chromatographic fingerprinting and quantitative analysis for the quality evaluation of Xinkeshu tablet." *Journal of Pharmaceutical Analysis* 2, no. 6 (2012): 422–430.

Weerakoon, S. W., L. S. R. Arambewela, G. A. S. Premakumara, and W. D. Ratnasooriya. "Note sedative activity of the crude extract of Rauvolfia densiflora." *Pharmaceutical Biology* 36, no. 5 (1998): 360–361.

Wiart, Christophe. *Medicinal plants of Asia and the Pacific*; Boca Raton: CRC Press, 2006.

Wilkins, Robert W., and Walter E. Judson. "The use of Rauwolif serpentina in hypertensive patients." *New England Journal of Medicine* 248, no. 2 (1953): 48–53.

Wink, M., T. Schmeller, and B. Latz-Brüning. "Modes of action of allelochemical alkaloids: interaction with neuroreceptors, DNA, and other molecular targets." *Journal of Chemical Ecology* 24, no. 11 (1998): 1881–1937.

Wink, Michael. "Modes of action of herbal medicines and plant secondary metabolites." *Medicines* 2, no. 3 (2015): 251–286.

Woodson, R. E., H. W. Yovngeen, E. Schlittler, and J. A. Schneider. *Rauwolfia: Botany, Pharmacognosy, Chemistry and Pharmacology*; Boston: Little, Brown and Company, 1957: pp. 32–33.

World Health Organization (WHO). *WHO Monographs on Selected Medicinal Plants*, Vol. 2; World Health Organization, 1999.

World Health Organization (WHO). "WHO Guide lines on Safety Monitoring of Herbal Medicines in Pharmacovigilance Systems." (2004).

World Health Organization (WHO). "Traditional medicine report by the secretariat eb 134/24." (2013). www.apps.who.int/gb/ebwha/pdf_files/EB134/B134_24-en.pdf. Accessed April 11, 2019.

Yadav, N. P., and V. K. Dixit. "Recent approaches in herbal drug standardization." *International Journal of Integrative Biology* 2, no. 3 (2008): 195–203.

Youngken, Heber W. "Malabar Rauwolfia, Rauwolfia micrantha Hook. f." *Journal of the American Pharmaceutical Association* 43, no. 3 (1954): 141–143.

Yu, Jun, and Qi Chen. "Antitumor activities of Rauwolfia vomitoria extract and potentiation of gemcitabine effects against pancreatic cancer." *Integrative Cancer Therapies* 13, no. 3 (2014): 217–225.

Yu, Jun, Yan Ma, Jeanne Drisko, and Qi Chen. "Antitumor activities of Rauwolfia vomitoria extract and potentiation of carboplatin effects against ovarian cancer." *Current Therapeutic Research* 75 (2013): 8–14.

Zeng, Jun, Dong-Bo Zhang, Pan-Pan Zhou, Qi-Li Zhang, Lei Zhao, Jian-Jun Chen, and Kun Gao. "Rauvomines A and B, two monoterpenoid indole alkaloids from Rauvolfia vomitoria." *Organic Letters* 19, no. 15 (2017): 3998–4001.

Zhang, H. Y., Y. Q. Gong, C. T. Cai, and G. X. Rao. "HPLC determination of reserpine in *Rauvolfia* plants." *Journal of Yunnan College of Traditional Chinese Medicine* 30 (2007): 7–9.

Zhang, Ying, Zhiqiang Huang, Li Ding, Hongfei Yan, Meiling Wang, and Shaohua Zhu. "Simultaneous determination of yohimbine, sildenafil, vardenafil and tadalafil in dietary supplements using high-performance liquid chromatography-tandem mass spectrometry." *Journal of Separation Science* 33, no. 14 (2010): 2109–2114.

Zirihi, G. N., Koffi N'Guessan, D. T. Etien, and B. Seri-Kouassi. "Evaluation in vitro of antiplasmodial activity of ethanolic extracts of Funtumia elastica, Rauvolfia vomitoria and Zanthoxylum gilletii on Plasmodium falciparum isolates from Côte-d'Ivoire." *Journal of Animal and Plant Sciences (JAPS)* 5, no. 1 (2009): 406–413.

Zysk, Kenneth Gregory. *Siddha Medicine in Tamil Nadu*; København: Nationalmuseet, 2008.

Index